**Finite State
Markovian Decision
Processes**

This is Volume 67 in
MATHEMATICS IN SCIENCE AND ENGINEERING
A series of monographs and textbooks
Edited by **RICHARD BELLMAN**, *University of Southern California*

A complete list of the books in this series appears at the end of this volume.

FINITE STATE MARKOVIAN DECISION PROCESSES

CYRUS DERMAN

DIVISION OF MATHEMATICAL METHODS
OF ENGINEERING AND OPERATIONS
RESEARCH
SCHOOL OF ENGINEERING AND
APPLIED SCIENCE
COLUMBIA UNIVERSITY
NEW YORK, NEW YORK

1970

ACADEMIC PRESS   New York and London

COPYRIGHT © 1970, BY ACADEMIC PRESS, INC.
ALL RIGHTS RESERVED
NO PART OF THIS BOOK MAY BE REPRODUCED IN ANY FORM,
BY PHOTOSTAT, MICROFILM, RETRIEVAL SYSTEM, OR ANY
OTHER MEANS, WITHOUT WRITTEN PERMISSION FROM
THE PUBLISHERS.

ACADEMIC PRESS, INC.
111 Fifth Avenue, New York, New York 10003

United Kingdom Edition published by
ACADEMIC PRESS, INC. (LONDON) LTD.
Berkeley Square House, London W1X 6BA

LIBRARY OF CONGRESS CATALOG CARD NUMBER: 70-117083

PRINTED IN THE UNITED STATES OF AMERICA

TO MY PARENTS

Samuel and Bessie (Segal) Derman

Contents

Preface   xi
Acknowledgments   xiii

1. Introduction

Preliminary Remarks	1
The Markovian Decision Model	2
Problems to Be Treated	5
Hierarchical Classification of Policies	6
Bibliographical Remarks	7
Problems	8

2. Finite Horizon Expected Cost Minimization

Dynamic Programming	11
Computational Example	15
Pathologies	16
Bibliographical Remarks	17
Problem	17

3. Some Existence Theorems

Summary	19
Expected Discounted Cost Problem	20
Expected Average Cost Problem	25
First-Passage Problem	28
First Passage as a Discounted Cost Problem	31
Bibliographical Remarks	32
Problems	33

4. Computational Methods for the Discounted Cost Problem

Introduction and Summary	35
Method of Successive Approximations	36
Policy Improvement Procedure	39
Linear Programming	41
Computational Examples	46
Bibliographical Remarks	49
Problems	50

5. Computational Procedures for the Optimal First-Passage Problem

Introduction	53
Method of Successive Approximations	54
Policy Improvement Procedure	56
Linear Programming Formulations	57
Computational Examples	59
The Finite Horizon Problem as a First-Passage Problem	61
Bibliographical Remarks	62
Problems	63

6. Expected Average Cost Criterion Computational Procedures

Summary	65
Policy Improvement Procedure	66
Linear Programming Formulations	73
Policy Improvement, Linear Programming under Irreducibility Assumption	78
Computational Example	81
Bibliographical Remarks	83
Problems	84

Contents ix

7. State-Action Frequencies and Problems with Constraints

Introduction and Summary	87
Expected State-Action Frequencies	88
Some Examples	89
The Main Theorems and Applications	91
State-Action Frequencies	98
Bibliographical Remarks	102

8. Optimal Stopping of a Markov Chain

Statement of the Problem	103
Stopping Problem as an Expected Average Gain Problem	104
A Different Approach	105
Computational Example	112
The Dual Linear Programming Problem	113
Some Other Forms of the Stopping Problem	114
Bibliographical Remarks	116
Problems	117

9. Some Applications

A Replacement Model	121
A Surveillance-Maintenance-Replacement Model	125
AOQL of Continuous Sampling Plans	130
A Sequential Search Problem	132
A Stochastic Traveling Salesman Problem	135
Bibliographical Remarks	137

Appendix A. **Markov Chains**	139
Bibliographical Notes	142
Appendix B. **Some Theorems from Analysis and Probability Theory**	143
Bibliographical Notes	147

Appendix C. **Convex Sets and Linear Programming**	149
Bibliographical Notes	152
References	153
Index	157

Preface

We are concerned with the optimal sequential control of certain types of dynamic systems. We assume such a system is observed periodically. After each observation the system is classified into one of a possible number of states; after each classification one of a possible number of decisions is implemented. The sequence of implemented decisions interacts with the chance environment to effect the evolution of the system. We call the mathematical abstraction of this process a Markovian decision process; however, some authors use the term discrete dynamic programming. Just as linear programming provides a general framework for formulating and solving certain optimization problems, so does the Markovian decision process provide a structure within which optimal control of dynamic systems can be formulated and solved.

Recognizing its potential usefulness we presented a course on Markovian decision processes in 1966 and 1968 for operations research students of the Columbia University School of Engineering and Applied Science. The lectures for the course have served as a basis for the preparation of this monograph.

This book is intended for operations researchers, statisticians, mathematicians, and engineers interested in mathematical methods for the control of dynamic systems. It can serve as a text for a course on dynamic programming which is intended to provide students with the basic computational algorithms as well as to prepare them for research in the subject.

Prerequisites include a reasonable grounding in real analysis or advanced calculus, a knowledge of the elementary theory of Markov chains, and an acquaintance with the rudiments of linear programming. An appendix on these prerequisites is provided; however, its primary purpose is to collect those facts which are explicitly used in the text and is not intended to serve as a source for obtaining the necessary background for reading this book.

Acknowledgments

Acknowledgments are due to Richard Bellman for his initial and recurrent encouragement to write this book; to Arthur Veinott for a reading of the manuscript and helpful comments; to, at least, Edward Ignall, Morton Klein, Peter Kolesar, and Howard Taylor for enlightening conversations; and to Regina Tetens and Stennett Parris for preparation of the manuscript. This work was supported, in part, by the Army, Navy, Air Force, and NASA under Contract N000 14-67-A-0108-0008 with the Office of Naval Research.

I

Introduction

Preliminary Remarks

Markovian decision processes are stochastic processes that describe the evolution of dynamic systems controlled by sequences of decisions or actions. Thus, in this monograph we shall be concerned with certain types of dynamic systems which are observed periodically, and influenced at the time of observation by the taking of one of a possible number of different actions. The evolution of the system will be the result of the interaction between the "laws of motion" of the system and the sequence of actions taken over time. The different paths of the system will have associated economic consequences; the ultimate aim is to take those actions that control the system in an optimal manner. Optimality will be defined relative to a stipulated criterion.

A typical system is an inventory system for a given product where the inventory level is under periodic review. After each review, the action taken is that of adding a certain amount of the product to the inventory level. The laws of motion of the system are determined by the pattern of demand for the product between times of review. Various costs associated with ordering new product, holding inventory in storage, shortages, etc. contribute to the economic consequences of the actions taken at the various times. A criterion for optimality would ordinarily be a function of long term costs.

Another typical system might consist of a component or group of components that is under periodic surveillance and subjected to periodic maintenance or replacement of one or more components. At each inspection the system is classified in some appropriate way and a decision is made as to what degree of maintenance to employ. The properties of the system together with the demands upon it determine the laws of motion. Economic aspects involve the various costs associated with maintenance and also the attributed costs due to failure of the system. Occasionally, failure costs may be difficult or impossible to ascertain, in which case "system reliability" may be a more appropriate yardstick with which to measure the effectiveness of a surveillance and maintenance procedure.

The Markovian Decision Model

To introduce the general model, let us assume that at points of time $t = 0, 1, \ldots$ the system is observed and classified into one of a possible number of states. We let $\{Y_t, t = 0, 1, \ldots\}$ denote the sequence of observed states. The letter I will denote the space of possible states. Throughout this volume, I will be a finite space.

After each observation of the system, one of a possible number of

The Markovian Decision Model 3

actions is taken. We let $\{A_t, t = 0, 1, \ldots\}$ denote the sequence of actions. K_i denotes the number of actions possible when the system is in state i. More frequently, we shall also use K_i to denote the set of possible actions when the system is in state i. No confusion should result from the double use of the notation. Throughout, K_i will be finite.

A *rule* or *policy*, to be denoted by R, is a prescription for taking actions at each point in time. We shall permit a policy for taking an action at time t to be a function of the entire "history" of the system up to time t. We will allow actions to be taken which are determined by a random mechanism, the random mechanism will be a function of the "history." For example, when in state i, a coin may be tossed to determine which of two possible actions to take. However, the kind of coin used may depend on the previous sequence of states and actions taken. In this volume the use of policies employing randomization enables the use of linear programming formulations of the problems of interest and allows one to obtain optimal policies in the face of certain constraints.

Thus, a policy R is a set of functions

$$\{D_a(H_{t-1}, Y_t), \quad a \in K_{Y_t}, \quad t = 0, 1, \ldots\}$$

satisfying

$$0 \leq D_a \leq 1 \quad \text{and} \quad \sum_a D_a = 1,$$

where H_t denotes the history of the system up to time t; that is, $H_t = \{Y_0, A_0, \ldots, Y_t, A_t\}$. One interprets $D_a(H_{t-1}, Y_t)$ as follows: if H_{t-1} denotes the history of the system up to time $t - 1$ and Y_t the state of the system at time t, then a random mechanism is to be used which assigns the probability $D_a(H_{t-1}, Y_t)$ of taking action a at time t.

We assume throughout that the laws of motion of the system can be characterized by a time invariant set of transition probabilities. Namely, whenever the system is in state i and action a is taken, then, regardless of its history, $q_{ij}(a)$ denotes the probability of the system being in state j at the next instant the system is observed. An alternative way of stating this assumption is that no matter what policy R is employed, the conditional probability that $Y_{t+1} = j$, given that H_{t-1}, $Y_t = i$, and $A_t = a$

is equal to $q_{ij}(a)$. In this volume it will be assumed that the set of numbers $\{q_{ij}(a), a \in K_i, i \in I, j \in I\}$ are known and, of course, satisfy

$$0 \leq q_{ij} \leq 1 \quad \text{and} \quad \sum_j q_{ij} = 1.$$

For example, consider the laws of motion of an inventory system under periodic review. Let Y_t denote the level of inventory at time t and let A_t denote the amount ordered after observing Y_t. Assume delivery of the A_t units is instantaneous so that at the moment of ordering, the inventory level is $Y_t + A_t$. Suppose the sequence of demands $\{D_t\}$ for the product during each of the periods is a sequence of independent and identically distributed random variables. If, for simplicity, we allow negative inventory (that is, backlogging of demand) and a denumerable state space, then

$$q_{ij}(a) = P\{Y_{t+1} = j \mid H_{t-1}, \quad Y_t = i, \quad A_t = a\}$$
$$= P\{\text{Demand} = i + a - j\}.$$

Given a distribution $P\{Y_0 = i\}$ over the initial states of the system and a policy R, then the sequence $\{Y_t, A_t, t = 0, 1, \ldots\}$ is a stochastic process. We call this process a *Markovian decision process*. The term Markovian is employed because of the special assumptions regarding the laws of motion. However, we point out that the process $\{Y_t, A_t, t = 0, 1, \ldots\}$ is not necessarily a Markov process. Since a policy R may be such that the prescription for taking actions is dependent upon the entire history of the process, the Markov property may not be satisfied by $\{Y_t, \mathbf{A}_t\}$.

In order to indicate the dependence of the probabilities on the policy R, the notation $P_R\{E\}$ will denote the probability of an event E occurring when policy R is used. The probability of the event E given the initial state $Y_0 = i$ and the use of the policy R will be denoted by $P_R\{E \mid Y_0 = i\}$.

We assume a certain cost structure superimposed on the Markovian decision process. Whenever the system is in state i and action a is taken, we assume that a known cost w_{ia} is incurred. For most of that considered in this volume, w_{ia} may also denote an expected cost rather than an actual cost. However, the important aspect of the assumption is that this cost is a function only of the state and action taken. For example, in

Problems to Be Treated

our inventory problem the cost incurred in a period is a function of the ordering costs and the inventory level at the end of that period. The expected value of this function taken with respect to the distribution of demand will yield our assumed cost w_{ia}, the expected cost associated with inventory level i, and the action of ordering a units.

We define the random variables $\{W_t, t = 0, 1, \ldots\}$:

$$W_t = w_{ia} \quad \text{if} \quad Y_t = i, \quad A_t = a, \quad a \in K_i, \quad i \in I.$$

We can then speak of expected costs; that is,

$$E_R W_t = \sum_i \sum_a P_R\{Y_t = i, \ A_t = a\} w_{ia}.$$

Since I and K_i will be finite, no question of existence of $E_R W_t$ will arise.

Problems to Be Treated

In terms of our model, we are now in the position to state some of the problems of interest in this volume. In each of the problems, it will be assumed that the initial state $Y_0 = i$ is given; that is, $P\{Y_0 = i\} = 1$.
Let

$$S_{R,T}(i) = E_R \sum_{t=0}^{T} W_t = \sum_{t=0}^{T} \sum_i \sum_a P_R\{Y_t = j, \ A_t = a\} w_{ja}.$$

In words, $S_{R,T}(i)$ is the expected total cost of operating the system up to and including the time "horizon" $t = T$, given that the initial state is i and the policy used is R. The problem of interest is that of obtaining that policy R which minimizes $S_{R,T}(i)$. This is the type of problem which is commonly dealt with by the straightforward method of dynamic programming. We discuss its solution in the next chapter.

Another problem treated is that of finding R to minimize $\sigma_R(i) = S_{R,\tau}(i)$, where τ denotes the smallest positive value of t such that $Y_t = j$, j being a "target" state at which the process is stopped. We refer to this as an optimal first-passage problem. It should be noted that τ is a random variable so that $\sigma_R(i)$ is the expected value of a random sum of random variables. For an obvious generalization, j need not refer to a single state, but may be a class of states.

A third problem to be dealt with is that involving the discounted cost criterion

$$\Psi_R(i, \alpha) = E_R \sum_{t=0}^{\infty} \alpha^t W_t,$$

where $0 \leq \alpha < 1$ (the discount factor) is a given number. We shall be interested in finding R to minimize $\Psi_R(i, \alpha)$.

A fourth problem arises from the expected average cost per unit time criterion:

$$\phi_R(i) = \lim_{T \to \infty} \frac{S_{R,T}(i)}{T+1}.$$

For some policies R, the limit of the above expression may not exist. In those cases we shall deal with the upper or lower limit, whichever seems appropriate. We shall be interested in finding a policy R to minimize $\phi_R(i)$.

Several other problems will be treated as well. We may have other criteria to minimize or we might be interested in minimizing $\phi_R(i)$ subject to certain side constraints. In Chapter 7 we hope to develop the theory so that these problems can be dealt with; in Chapter 8 a stopping problem is considered.

Hierarchical Classification of Policies

In each case we shall be concerned with three questions: existence, structure, and computational procedures. With regard to each of these it is convenient to introduce a hierarchical classification of the totality of possible policies. We let C denote the class of *all* policies under consideration; that is, those with possible dependency on the complete history of the system. We let C_M denote the class of all memoryless or Markovian type policies. That is, C_M consists of all policies R such that $D_a(H_{t-1}, Y_t)$ is a function only of Y_t, t, and a. When $R \in C_M$, then $\{Y_t, t = 0, 1, \ldots\}$ is a Markov chain, not necessarily stationary. We let

C_S denote the class of all Markovian policies which are time invariant. That is, C_S consists of all policies R such that $D_a(H_{t-1}, Y_t)$ are functions only of Y_t and a. Let $D_{ia} = D_a\{H_{t-1}, Y_t = i\}$ when $R \in C_s$. Then $\{Y_t, t = 0, 1, \ldots\}$ is a Markov chain with stationary transition probabilities

$$P_{ij} = \sum_a D_{ia} q_{ij}(a), \qquad i, j \in I.$$

Finally, we let C_D denote the subclass of C_S consisting of the deterministic policies. That is, $R \in C_D$ whenever D_{ia} is 0 or 1 for every $i \in I$. In this case we can think of R as defining a single-valued transformation from the states to the actions; that is, when $R \in C_D$, to each state i there corresponds an action a_i, among the possible actions K_i, such that R prescribes action a_i when the system is in state i. Accordingly, when convenient and $R \in C_D$, we shall employ the notation w_{iR} and $q_{ij}(R)$ to denote w_{ia_i} and $q_{ij}(a_i)$.

Since the class C of all policies is infinite, the question of existence of an optimal policy will be important for each problem considered. In all problems dealt with here, we shall want to assert that not only does an optimal policy exist but that one is also a member of C_M, C_S, or C_D. In other words, we shall want to say something about the structure of at least one of the optimal policies. In certain special cases, perhaps, more can be said about structure. When the structure is such that an optimal policy is a member of C_M, C_S, or C_D, then frequently, a computational procedure can be obtained for its determination.

Bibliographical Remarks

The title of this book might well have been called "Dynamic Programming," or better, "Discrete Dynamic Programming" as used by Blackwell [6]. The descriptive phrase "Markovian Decision Process" is due to Bellman [2], and because of the connections of the material treated herein with Markov chains, we prefer the latter description.

The Markovian decision model in recent years has been the subject of an increased amount of research activity. Early papers, in a special context by Bellman and LaSalle [4], Bellman and Blackwell [5], and later, more generally by Shapley [48] were among the first formulations of the model in the context of two-person dynamic games. Its first explicit formulation outside the game context is given by Bellman [2]. The model has a large number of parameters and is readily adaptable to many dynamic systems as a descriptive model. Although some of the methods of dynamic programming such as backward induction and method of successive approximations predate the formal conception of dynamic programming and, in particular, the Markovian decision model, it was not until computational breakthroughs by Howard [34], Manne [46], and D'Epenoux [14], some seven years or so after Shapley's [48] treatment, that interest in the model increased and an awareness in its potential usefulness developed.

In the Shapley [48] two-person stochastic games model, as in the earlier papers by Bellman and LaSalle [4], and Bellman and Blackwell [5], the process $\{Y_t, t = 0, 1, \ldots\}$ is controlled by two sets of simultaneous action. The "laws of motion" are in the form of numbers $\{q_{ij}(a, b), a \in K_i^{\mathrm{I}}, b \in K_i^{\mathrm{II}}, i \in I, j \in I\}$, where K_i^{I} and K_i^{II}, $i \in I$, are sets of possible actions for a "player I" and a "player II" at state i. That is, if the process is in state i and player I takes action a and player II takes action b then the probability is $q_{ij}(a, b)$ that the next period will find the process in state j. The costs in this model are of the form w_{iab}, to be interpreted as the cost to player I and the gain to player II when the process is in state i and player I takes action a and player II takes action b. Thus, the process under consideration in this volume concerns the special case where player II has only one available action at each state.

Problems

(1) Derive the form of the laws of motion, the $q_{ij}(a)$'s, for the example of the inventory system under periodic review when backlogging is not permitted.

Problems 9

(2) For an inventory system under periodic review for the cases of backlogging and no backlogging, construct the forms of the costs $\{w_{ia}\}$.

(3) For the inventory system under periodic review construct a policy R that belongs to C_D; to $C_S - C_D$; to $C - C_S$.

2

Finite Horizon Expected Cost Minimization

Dynamic Programming

This chapter is concerned with the determination of that policy $R \in C$ which minimizes $S_{R,T}(i) = E_R \sum_{i=0}^{T} W_t$, where the horizon T and initial state i are given. We will show that a backward induction method, which is the essence of dynamic programming, provides a computational procedure for obtaining the optimal policy. Although I is assumed to be finite, the method of this chapter holds for countable I as long as the costs $\{w_{ia}\}$ are such that $E_R W_t$ is well defined for all R and t.

Let us denote by $V_n(R, j, h_{n-1})$, $0 \leq n \leq T$, the conditional expected total cost of a process from time $t = n$ to time $t = T$ given the history $H_{n-1} = h_{n-1}$, $Y_n = j$ and policy R; that is,

$$V_n(R, j, h_{n-1}) = E_R\left\{\sum_{t=n}^{T} W_t \mid Y_n = j, \; H_{n-1} = h_{n-1}\right\}, \quad 0 \leq n \leq T.$$

When $n = 0$, we have, since there is no history,

$$V_0(R, i) = S_{R,T}(i).$$

Set

$$V_n^*(i, h_{n-1}) = \inf_{R \in C} V_n(R, i, h_{n-1}), \quad 0 \leq n \leq T.$$

Then providing $V_0^*(i) = \min_{R \in C} V_0(R, i)$, the value of $S_{R,T}(i)$ is $V_0^*(i)$ when an optimal policy is used. It will be seen that the minimum over all $R \in C$ is obtained. The method of backward induction employs a recursion formula by which V_{n-1}^* can be expressed in terms of V_n^* for $n = T, T-1, \ldots, 1$, thereby achieving the value $V_0^*(i)$. At the same time the optimal policy is perceived.

We first prove:

LEMMA 1. *For every* $H_{n-1} = h_{n-1}$, $V_n^*(i, h_{n-1}) = V_n^*(i)$, $n = 1, \ldots, T, i \in I$; *that is to say*, $V_n^*(i, h_{n-1})$ *is independent of* h_{n-1}.

Proof: Fix i. Let $v^*(i) = \inf_{R, h_{n-1}} V_n(R, i, h_{n-1})$. Let $\varepsilon > 0$ be given arbitrarily. Let R_0, h_{n-1}^0 be such that

$$V_n(R_0, i, h_{n-1}^0) < v^*(i) + \varepsilon.$$

Define R_1 as follows:

$$D_a^{R_1}(H_{t-1}, Y_t) = D_a^{R_0}(H_{t-1}, Y_t), \quad t = 0, \ldots, n-1,$$
$$= D_a^{R_0}(h_{n-1}^0, Y_n, \ldots, Y_t), \quad t = n, \ldots, T;$$

that is, R_1 is the same as R_0 for $t = 0, \ldots, n-1$, but, for $t = n, \ldots, T$,

Dynamic Programming

R_1 prescribes actions as if policy R_0 were in effect and the history $H_{n-1} = h_{n-1}^0$ had been observed up to time $n-1$. Then, for every $H_{n-1} = h_{n-1}$,

$$V_n(R_1, i, h_{n-1}) = V_n(R_0, i, h_{n-1}^0) \leq v^*(i) + \varepsilon.$$

Therefore, since $\varepsilon > 0$ is arbitrary, $V_n^*(i, h_{n-1}) \leq v^*(i)$. On the other hand, by definition of $v^*(i)$, we have that

$$V_n(R, i, h_{n-1}) \geq v^*(i)$$

for every R and h_{n-1}. Hence, $V_n^*(i, h_{n-1}) = v^*(i)$ independent of h_{n-1} and the lemma is proved.

For the following theorem, we set $V_{T+1}^*(i) = 0, i \in I$. We prove:

THEOREM 1. *If R^* is defined as a policy which at time n takes action a_i^* (a function of n) satisfying*

$$w_{ia^*_i} + \sum_j q_{ij}(a_i^*) V_{n+1}^*(j) = \min_a \left\{ w_{ia} + \sum_j q_{ij}(a) V_{n+1}^*(j) \right\}$$

for $i \in I$ and $n = 0, \ldots, T$, then $V_n(R^*, i, H_{n-1}) = V_n^*(i), n = 0, 1, \ldots, T, i \in I$. In particular, R^* is optimal for minimizing $S_{R,T}(i)$.

Proof: We will use backward induction on n. Suppose $n = T$. Then for any R and h_{T-1}

$$E_R\{W_T | Y_T = i, H_{T-1} = h_{T-1}\} = \sum_a D_a^R(h_{T-1}, i) w_{ia}$$

$$\geq \min_a \{w_{ia}\}$$

$$= \min_a \left\{ w_{ia} + \sum_j q_{ij}(a) V_{T+1}^*(j) \right\}$$

$$= V_T(R^*, i, h_{T-1}).$$

Since, in fact, $V_T(R^*, i, h_{T-1})$ is independent of h_{T-1}, we have that it is equal to $V_T^*(i)$. Now assume that $V_t(R^*, i, h_{t-1}) = V_t^*(i)$ for $t = n+1, \ldots, T$. We shall show that the same holds for $t = n$. For any R and h_{n-1},

$$V_n(R, i, h_{n-1}) = E_R\left\{\sum_{t=n}^{T} W_t | Y_n = i, H_{n-1} = h_{n-1}\right\}$$

$$= \sum_a D_a^R(h_{n-1}, i)w_{ia} + \sum_j \sum_a D_a^R(h_{n-1}, i)q_{ij}(a)$$

$$\times E_R\left\{\sum_{t=n+1}^{T} W_t | Y_{n+1} = j, Y_n = i, A_n = a, h_{n-1}\right\}$$

$$\geq \sum_a D_a^R(h_{n-1}, i)\left\{w_{ia} + \sum_j q_{ij}(a)V_{n+1}^*(j)\right\}$$

$$\geq \min_a\left\{w_{ia} + \sum_j q_{ij}(a)V_{n+1}^*(j)\right\}$$

$$= V_n(R^*, i, h_{n-1}).$$

The first inequality follows from Lemma 1 and the last equation follows from the induction assumption. The right-hand side is again independent of h_{n-1}. Hence, $V_n(R^*, i, h_{n-1}) = V_n^*(i)$, $i \in I$. This proves the theorem.

COROLLARY 1. $V_n^*(i) = \min_a\{w_{ia} + q_{ij}(a)V_{n+1}^*(j)\}$, $i \in I$, $n = 0, 1, \ldots, T$.

Proof: The equations follow from the fact that $V_n^*(i) = V_n(R^*, i, h_{n-1})$, for $i \in I$ and $n = 0, 1, \ldots, T$ and the last equality of the proof of the theorem.

COROLLARY 2. R^* is a member of C_M.

Proof: This is apparent from the definition of R^*.

The defining equations of R^* of Corollary 1 are known as the functional equations of dynamic programming. They provide a simple but extremely useful recursive scheme for obtaining the optimal policy as long as the state space I is not too large. They also express what is commonly referred to as the "principle of optimality" which asserts that an optimal policy for minimizing $S_{R,T}(i)$ must also minimize $V_n(R, i, h_{n-1})$ for every $n = 0, 1, \ldots, T$.

Computational Example

Suppose $I = \{0, 1\}$; $K_i = 2$, $i = 0, 1$ with

$$\begin{pmatrix} w_{01} & w_{02} \\ w_{11} & w_{12} \end{pmatrix} = \begin{pmatrix} 1 & 0 \\ 2 & 2 \end{pmatrix},$$

and

$$\begin{pmatrix} (q_{00}(1), q_{00}(2)) & (q_{01}(1), q_{01}(2)) \\ (q_{10}(1), q_{10}(2)) & (q_{11}(1), q_{11}(2)) \end{pmatrix} = \begin{pmatrix} (\tfrac{1}{2}, \tfrac{1}{4}) & (\tfrac{1}{2}, \tfrac{3}{4}) \\ (\tfrac{2}{3}, \tfrac{1}{3}) & (\tfrac{1}{3}, \tfrac{2}{3}) \end{pmatrix}.$$

We want to find R to minimize $S_{R,T}(i)$, $(i = 0, 1)$ for $T = 2$. First, we calculate $V_2^*(i)$, $(i = 0, 1)$:

$$V_2^*(0) = \min\{w_{01}, w_{02}\} = 0,$$
$$V_2^*(1) = \min\{w_{11}, w_{12}\} = 2,$$

keeping in mind that $a_0^*(2)$ (the optimal action taken at $t = 2$ when in state 0) is 2 and that $a_1^*(2) = 1$ or 2. Now

$$V_1^*(0) = \min(1 + \tfrac{1}{2}V_2^*(0) + \tfrac{1}{2}V_2^*(1), 3 + \tfrac{1}{4}V_2^*(0) + \tfrac{3}{4}V_2^*(1))$$
$$= \min(1 + 1, 3 + \tfrac{3}{2})$$
$$= 2;$$
$$V_1^*(1) = \min(2 + \tfrac{2}{3}V_2^*(0) + \tfrac{1}{3}V_2^*(1), 1 + \tfrac{1}{3}V_2^*(0) + \tfrac{2}{3}V_2^*(1))$$
$$= \min(2 + \tfrac{2}{3}, 1 + \tfrac{4}{3})$$
$$= \tfrac{7}{3}.$$

Here $a_0^*(1) = 1$, $a_1^*(1) = 2$. Then

$$V_0^*(0) = \min(1 + \tfrac{1}{2}V_1^*(0) + \tfrac{1}{2}V_2^*(0), 3 + \tfrac{1}{4}V_1^*(0) + \tfrac{3}{4}V_1^*(1))$$
$$= \min(1 + \tfrac{13}{6}, 3 + \tfrac{27}{12})$$
$$= \tfrac{19}{6};$$
$$V_0^*(1) = \min(2 + \tfrac{2}{3}V_1^*(0) + \tfrac{1}{3}V_1^*(1), 1 + \tfrac{1}{3}V_1^*(0) + \tfrac{2}{3}V_1^*(1))$$
$$= \min(2 + \tfrac{19}{9}, 1 + \tfrac{20}{9})$$
$$= \tfrac{29}{9},$$

with $a_0^*(0) = 1$, $a_1^*(0) = 2$.

Therefore, the optimal policy with respect to minimizing $S_{R,2}(i)$ is to take actions 1, 1, and 2 when in state 0 at times $t = 0, 1, 2$, respectively, and to take actions 2, 2, and 1 or 2 when in state 1 at times $t = 0, 1, 2$, respectively.

Pathologies

When the state and action spaces are noncountably infinite, one may encounter technical difficulties which restrict the universal validity of the backward induction method. These difficulties arise in pathological cases where not *all* policies possess an associated $E_R W_t$ defined for all t; only those policies which satisfy certain measurability conditions can be so evaluated. Thus, the use of the criterion of mathematical expectation or, indeed, the assumption that $\{Y_t\}$ be a stochastic process has the effect of imposing subtle constraints on the space of possible policies, which in turn raises the possibility that the functional equation approach has some flaws. In order to appreciate how constraints on acceptable policies can invalidate the dynamic programming procedure, even in the finite state and action case, it should be noted that the imposition of the gross restriction that R be a member of C_D nullifies the fact that the optimal policy will satisfy the functional equations.

Bibliographical Remarks

For earlier and fuller expositions of the material of this chapter one should refer to Bellman [3]. So intuitive is the dynamic programming or backward induction approach, that one rarely encounters a formal proof that the method yields an optimal policy; hence, the proof is given here, despite the fact that for discrete state spaces and finite actions at each state the procedure is clearly correct. The pathology alluded to when state and action spaces are noncountable was revealed by Blackwell [7].

Problem

(1) Suppose $I = \{0, 1, 2\}$; $K_i = 2$, $i = 0, 1, 2$, where

$$\begin{pmatrix} w_{01} & w_{02} \\ w_{11} & w_{12} \\ w_{21} & w_{22} \end{pmatrix} = \begin{pmatrix} 1 & 0 \\ 2 & 1 \\ 1 & 2 \end{pmatrix}$$

and

$$\begin{pmatrix} (q_{00}(1), q_{00}(2)) & (q_{01}(1), q_{01}(2)) & (q_{02}(1), q_{02}(2)) \\ (q_{10}(1), q_{10}(2)) & (q_{11}(1), q_{11}(2)) & (q_{12}(1), q_{12}(2)) \\ (q_{20}(1), q_{20}(2)) & (q_{21}(1), q_{21}(2)) & (q_{22}(1), q_{22}(2)) \end{pmatrix}$$

$$= \begin{pmatrix} (\tfrac{1}{2}, \tfrac{1}{3}) & (\tfrac{1}{4}, \tfrac{1}{3}) & (\tfrac{1}{4}, \tfrac{1}{3}) \\ (0, \tfrac{1}{3}) & (1, \tfrac{2}{3}) & (0, 0) \\ (\tfrac{2}{3}, 0) & (0, \tfrac{1}{3}) & (\tfrac{1}{3}, \tfrac{2}{3}) \end{pmatrix}.$$

Find R^* for $T = 3$.

1

3

Some Existence Theorems

Summary

In this chapter we shall prove the existence of optimal policies in the class C_D for the expected discounted cost, expected average cost, and first-passage problems. The criterion in the expected average cost problem is at first $\phi_R(i) = \limsup S_{R,T}(i)/T + 1$. We then obtain the same result for the problem based on the lower limit definition of $\phi_R(i)$. Our method here involves discussion of the discounted cost problem first and then, using the results obtained together with elementary Abelian theorems, proving existence theorems for the other two problems.

Expected Discounted Cost Problem

Our basic approach to existence in the discounted case is to first establish that $\Psi_R(i, \alpha)$ for fixed i and α, $0 \leq \alpha < 1$, is a continuous function of R and that C is a compact space. Thus an $R \in C$ minimizing $\Psi_R(i, \alpha)$ must exist. From there, we establish that there exists an $R \in C_D$ minimizing $\Psi_R(i, \alpha)$.

We say a sequence $\{R_n, n = 1, 2, \ldots\}$ of policies converges to a policy R if for every a, y_t, h_t, $t = 0, 1, \ldots$, $\lim_{n \to \infty} D_a^{R_n}(h_t, y_t) = D_a^R(h_t, y_t)$.
We say a class of policies is compact if for every sequence of policies $\{R_n, n = 1, 2, \ldots\}$ there exists a subsequence $\{R_{n_k}, k = 1, 2, \ldots\}$ that converges to a policy in the class.

We first prove:

LEMMA 1. *The class C is compact.*

Proof: For every $H_t = h_t$, $Y_t = y_t$, the space $D(h_t, y_t) = \{D_1(h_t, y_t), \ldots, D_{K_{y_t}}(h_t, y_t)\}$ is compact since K_i is finite for every $i \in I$. By Tychonov's theorem (Theorem 3 of Appendix B), the product space $\prod_{h_t, y_t, t} D(h_t, y_t)$ is also compact. However, every point in this product space is by definition a policy and every policy corresponds to a point in the product space. Hence, the product space is the space C. Thus C is compact.

The temptation is to assert that Lemma 1 holds for any state space I and compact action spaces. However, because of the measurability constraints alluded to in Chapter 2, C, being the class of *all* policies for which stochastic processes $\{Y_t, A_t, t = 0, 1, \ldots\}$ and expectations

Expected Discounted Cost Problem

EW_t, $t = 0, 1, \ldots$ are well defined, may be a proper subclass of the product space $\prod_{h_t, y_t, t} D(h_t, y_t)$ and therefore, may not be compact.

LEMMA 2. Let t be arbitrary and $H_t = h_t = \{Y_0 = i, A_0 = a_0, \ldots, Y_t = i_t, A_t = a_t\}$ be given, then $P_R\{H_t = h_t \mid Y_0 = i\}$ is a continuous function of R.

Proof: For $t = 0$, $P_R\{H_0 = h_0 \mid Y_0 = i\} = D_{a_0}^R(Y_0 = i)$, and hence, the assertion is true for $t = 0$. Assume it true for $t = 0, \ldots, T$. Then since

$$P_R\{H_{T+1} = h_{T+1} \mid Y_0 = i\}$$
$$= P_R\{H_T = h_T \mid Y_0 = i\} q_{i_T, i_{T+1}}(a_T) D_{a_{T+1}}^R(h_T, Y_{T+1} = i_{T+1}).$$

we have by induction that the assertion is true for $t = T + 1$ and the lemma is proven.

LEMMA 3. Let t be arbitrary and $Y_t = j$, $A_t = a_t$ be given, then $P_R\{Y_t = j, A_t = a_t \mid Y_0 = i\}$ is a continuous function of R.

Proof: We can write

$$P_R\{Y_t = j, \quad A_t = t \mid Y_0 = i\}$$
$$= \sum_{h_{t-1}} P_R\{Y_t = j, \quad A_t = a_t \mid Y_0 = i, \quad H_{t-1} = h_{t-1}\}$$
$$\times P_R\{H_{t-1} = h_{t-1} \mid Y_0 = i\}$$
$$= \sum_{h_{t-1}} q_{i_{t-1}, j}(a_{t-1}) D_{a_t}^R(h_{t-1}, Y_t = j) \cdot P_R\{H_{t-1} = h_{t-1} \mid Y_0 = i\}.$$

Since there is only a finite number of h_{t-1}'s, Lemma 3 follows from Lemma 2.

We remark that if I is countable so that a countable number of histories h_t exist for each t, Lemma 3 can still be established.

LEMMA 4. $E_R W_t$, $t = 0, 1, \ldots$ and $\Psi_R(i, \alpha)$ for a given $0 \leq \alpha < 1$ are continuous functions of R.

Proof: For a given t, $E_R W_t$ is a finite linear combination of terms $P_R\{Y_t = j, A_t = a \mid Y_0 = i\}$. Thus from Lemma 3, $E_R W_t$ is continuous. Similarly $\Psi_R(i, T) = \sum_{t=0}^{T} \alpha^t E_R W_t$ is continuous for every $T = 0, 1, \ldots$. Since $\Psi_R(i, \alpha) = \lim_{T \to \infty} \Psi_R(i, \alpha, T)$ uniformly in R, $\Psi_R(i, \alpha)$ is also continuous.

We are now in a position to prove:

LEMMA 5. Let $Y_0 = i$ be given. There exists an $R^* \in C$ such that

$$\Psi_{R^*}(i, \alpha) = \inf_{R \in C} \Psi_R(i, \alpha), \qquad i \in I.$$

Proof: Let

$$\Psi_R(\alpha) = \sum_i \beta_i \Psi_R(i, \alpha),$$

where β_i, $i \in I$ are given positive numbers. Since by Lemma 1, C is compact and by Lemma 4, $\Psi_R(i, \alpha)$ is a continuous function of R, then $\Psi_R(\alpha)$ is also a continuous function of R and hence, from the well-known fact (Theorem 2 of Appendix B) that continuous functions over compact spaces achieve their extremes, $\Psi_R(\alpha)$ is minimized by a policy $R^* \in C$. However, R^* must also minimize $\Psi_R(i, \alpha)$ for each $i \in I$, otherwise a different policy could easily be constructed which would yield a smaller value for $\Psi_R(\alpha)$.

Define a_i, $i \in I$, as those actions which satisfy

$$w_{ia_i} + \alpha \sum_j q_{ij}(a_i) \Psi_{R^*}(j, \alpha)$$
$$= \min_a \left\{ w_{ia} + \alpha \sum_j q_{ij}(a) \Psi_{R^*}(j, \alpha) \right\}, \qquad i \in I. \qquad (1)$$

Expected Discounted Cost Problem

If a_i is not uniquely defined, let it be any one of the several actions satisfying (1). Let R_0 be defined as that policy which takes action a_i when the system is in state i, $i \in I$. Here, R_0 is a member of C_D. We now prove

THEOREM 1. *The policy R_0 minimizes $\Psi_R(i, \alpha)$ for all $i \in I$, and $\Psi_{R_0}(i, \alpha)$ uniquely satisfies*

$$\Psi_{R_0}(i, \alpha) = \min_a \{w_{ia} + \alpha \sum q_{ij}(a) \Psi_{R_0}(j, \alpha)\}, \qquad i \in I. \qquad (2)$$

Proof: For each $n = 1, 2, \ldots$ let R_n be a policy defined as follows:

$$D_{a_i}^{R_n}(h_{t-1}, Y_t = i) = 1 \qquad \text{for every} \quad H_{t-1} = h_{t-1}$$
$$\text{and} \quad t = 0, 1, \ldots, n-1$$

and

$$D_a^{R_n}(H_{t-1}, Y_t) = D_a^{R^*}(H_{t-n}^{(n)}, Y_{t-n}^{(n)}) \qquad \text{for} \quad t \geq n,$$

where

$$Y_{t-n}^{(n)} = Y_t, \qquad A_{t-n}^{(n)} = A_t,$$
$$H_{t-n}^{(n)} = \{Y_0^{(n)}, A_0^{(n)}, \ldots, Y_{t-n}^{(n)}, A_{t-n}^{(n)}\}.$$

In words, for $t = 0, \ldots, n-1$, R_n prescribes action a_i whenever the system is in state i. Thereafter it prescribes according to R^* as if the process started anew at time $t = n$. Now, from Lemma 1 of Chapter 2,

$$E_R\left\{\sum_{t=1}^\infty \alpha^t W_t \mid Y_0 = i, A_0 = a, Y_1 = j\right\} \geq \inf_{R \in C} E_R\left\{\sum_{t=1}^\infty \alpha^t W_t \mid Y_1 = j\right\}$$

for all $R \in C$; therefore, for each $i \in I$ we have

$$\Psi_{R^*}(i, a) = \sum_a D_a^{R^*}(Y_0 = i) w_{ia} + \alpha \sum_j \sum_a D_a^{R^*}(Y_0 = i) q_{ij}(a)$$

$$\times E_{R^*}\left\{\sum_{t=1}^\infty \alpha^{t-1} W_t \mid Y_0 = i, A_0 = a, Y_1 = j\right\}$$

$$= \sum_a D_a^{R^*}(Y_0 = i)\left\{w_{ia} + \alpha \sum_j q_{ij}(a)\right.$$

(equation continued)

$$\times E_{R^*}\left(\sum_{t=1}^{\infty}\alpha^{t-1}W_t\mid Y_0=i,\ A_0=a,\ Y_1=j\right)\bigg\}$$

$$\geq \sum_a D_a^{R^*}(Y_0=i)\bigg\{w_{ia}+\alpha\sum_j q_{ij}(a)\inf_{R\in C}E_R\left(\sum_{t=1}^{\infty}\alpha^{t-1}W_t\mid Y_1=j\right)\bigg\}$$

$$= \sum_a D_a^{R^*}(Y_0=i)\bigg\{w_{ia}+\alpha\sum_j q_{ij}(a)\Psi_{R^*}(j,\alpha)\bigg\}$$

$$\geq \min_a\bigg\{w_{ia}+\alpha\sum_j q_{ij}(a)\Psi_{R^*}(j,\alpha)\bigg\}$$

$$= \Psi_{R_1}(i,\alpha).$$

Consequently, since $\Psi_{R^*}(i,\alpha)$ is minimal we have that the equality holds. By repeated iteration it follows that $\Psi_{R^*}(i,\alpha)=\Psi_{R_n}(i,\alpha)$ for $n=1,2,\ldots$ and all $i\in I$. Since $\Psi_R(i,\alpha)$ is a continuous function of R, and $\{R_n, n=1,2,\ldots\}$ converges to R_0, it follows that $\Psi_{R^*}(i,\alpha)=\Psi_{R_0}(i,\alpha), i\in I$. This establishes the optimality of R_0 and also that $\Psi_{R_0}(i,\alpha)$ satisfies (2). Uniqueness will be shown in Corollary 1 of Theorem 1 of Chapter 4.

COROLLARY 1. *There exists an $\tilde{R}\in C_D$ such that for each $i\in I$, $\Psi_{\tilde{R}}(i,\alpha)=\inf_{R\in C}\Psi_R(i,\alpha)$ for all α near enough to* 1.

Proof: From Theorem 1 we need only show that

$$\Psi_{\tilde{R}}(i,\alpha)=\inf_{R\in C_D}\Psi_R(i,\alpha)$$

for all α near enough to 1. However, for any $R\in C_D$, it is easily seen that

$$\Psi_R(i,\alpha)=w_{iR}+\alpha\sum_j q_{ij}(R)\Psi_R(j,\alpha),\qquad i\in I,$$

from which it follows (see Theorem 3 of Appendix A) that $\Psi_R(i,\alpha)$ is a rational function of α for $0\leq\alpha<1$. Let $\{\alpha_n, n=1,2,\ldots\}$ be a sequence such that $\lim_{n\to\infty}\alpha_n=1$ and $R_{\alpha_1}=R_{\alpha_2}=\cdots=\tilde{R}$ (say) where $R_{\alpha_n}\in C_D$ minimizes $\Psi_R(i,\alpha_n)$. Such a sequence can be chosen since C_D

Expected Average Cost Problem

is a finite set. Since the difference $\Psi_R(i, \alpha) - \Psi_{\bar{R}}(i, \alpha)$ is also rational it is either identically zero or has, at most, a finite number of zeros. Thus, there is an interval $(\alpha(R, i), 1)$ for which $\Psi_R(i, \alpha) - \Psi_{\bar{R}}(i, \alpha) \geqq 0$ for all $\alpha \in (\alpha(R, i), 1)$. Let $\tilde{\alpha} = \max_{R,i} \alpha(R, i)$. Then for $\alpha > \tilde{\alpha}$, $R \in C_D$, we have $\Psi_R(i, \alpha) \geqq \Psi_{\bar{R}}(i, \alpha)$ and the corollary is proved.

Expected Average Cost Problem

We now turn to the problem of finding $R \in C$ to minimize

$$\phi_R(i) = \lim_{T \to \infty} \sup S_{R,T}(i)/(T + 1)$$

the expected average cost per unit time over an infinite time horizon, given the initial state $Y_0 = i$. Since $\lim_{T \to \infty} S_{R,T}(i)/T + 1$ does not exist in general, we define $\phi_R(i)$ by the upper limit. However, Corollary 1 to Theorem 2 below will treat the case where $\phi_R(i)$ is defined by the lower limit. We prove:

THEOREM 2. There exists a policy $R^* \in C_D$ such that

$$\phi_{R^*}(i) = \inf_{R \in C} \phi_R(i), \qquad i \in I.$$

Proof: Let $R^* \in C_D$ be such that

$$\Psi_{R^*}(i, \alpha) = \inf_{R \in C} \Psi_R(i, \alpha), \qquad i \in I,$$

for every α near enough to 1. Corollary 1 to Theorem 1 guarantees the existence of such a policy R^*. We shall now show that R^* is an optimal policy for the criterion $\phi_R(i)$. From Theorem 1(b) of Appendix B, since $\phi_R(i) = \lim_{T \to \infty} S_{R,T}(i)/T + 1$ when $R \in C_D$ (a consequence of Theorem 1 of Appendix A), we have

$$\phi_{R^*}(i) = \lim_{\alpha \to 1}(1 - \alpha)\Psi_{R^*}(i, \alpha);$$

from Theorem 1(c) of Appendix B we have for all $R \in C$ that

$$\limsup_{\alpha \to 1}(1 - \alpha)\Psi_R(i, \alpha) \leq \limsup_{T \to \infty} \frac{S_{R,T}(i)}{T + 1}, \quad i \in I.$$

Using the fact that $\Psi_{R^*}(i, \alpha) \leq \Psi_R(i, \alpha)$ for all α near enough to 1 combined with the two above inequalities yields

$$\phi_{R^*}(i) \leq \limsup_{T \to \infty} \frac{S_{R,T}(i)}{T + 1}, \quad i \in I,$$

and the theorem is proved.

We remark that more than one optimal policy may exist. In our construction of the proof we showed that the policy R^*, which is optimal with respect to $\Psi_R(i, \alpha)$ for every α near enough to 1, is also optimal with respect to $\phi_R(i)$. However, not every policy that is optimal with respect to $\phi_R(i)$ will be optimal with respect to $\Psi_R(i, \alpha)$ for every α near enough to 1. The following example demonstrates this.

Let I consist of the states 0 and 1. Let $K_0 = 2$, $K_1 = 1$, where $q_{00}(1) = \beta > 0$, $q_{01}(2) = 1$, $q_{11}(1) = 1$, $w_{01} = 1$, $w_{02} = 0$, $w_{11} = 0$. Here, C_D contains two policies. Let R_1 denote the policy in C_D which takes action 1 in state 0 and R_2, which takes action 2 in state 0. Clearly, $\phi_{R_1}(0) = \phi_{R_2}(0)$. However, $\Psi_{R_1}(0, \alpha) = \sum_{t=0}^{\infty} (\alpha\beta)^t$, $\Psi_{R_2}(0, \alpha) = 0$. Thus R_2 is a better choice than R_1 with respect to the discounted cost criterion. For more on this subject see the bibliographical notes of Chapter 6.

Suppose we define $\phi_R(i) = \liminf_{T \to \infty} S_{R,T}(i)/(T + 1)$. As a consequence of Theorem 2 we have:

COROLLARY 1. There exists a policy R^* (the same policy as in Theorem 2) such that

$$\phi_{R^*}(i) = \inf_{R \in C} \phi_R(i), \quad i \in I,$$

when $\phi_R(i)$ is defined as $\liminf_{T \to \infty} S_{R,T}(i)/T + 1$.

Expected Average Cost Problem   27

Proof: If R is such that $\lim_{T \to \infty} S_{R,T}(i)/T + 1$ exists for every $i \in I$, then by Theorem 2 $\phi_{R*}(i) \leq \phi_R(i)$, $i \in I$. Suppose that for every i and j there exists a policy $R_{ji} \in C_D$ such that $P_{R_{ji}}\{Y_t = i$ for some $t \geq 1 \mid Y_0 = i\} = 1$, and m_{ji}, the mean first-passage time from j to i under R_{ji}, is finite. Starting with an arbitrary policy R, define $\tilde{R}$ as follows: use policy R for times $t = 0, \ldots, T$; then if $Y_0 = i$ and $Y_T = j$, use policy R_{ji} until $Y_t = i$ for the first time after T, after which use policy R as if starting from $t = 0$. Repeat this construction; that is, each time policy R_{ji}, depending on the state j, returns the process to state i, use policy R as if starting from $t = 0$ for $T + 1$ units of time, then use policy R_{ji} where j is the state of the process at the time of the switch in policies. The process $\{Y_t\}$ under $\tilde{R}$ is a recurrent event process (see Appendix B) and by Theorem 6, Appendix B, $\lim_{T \to \infty} S_{R,T}(i)/(T + 1)$ exists. Hence, $\phi_{R*}(i) \leq \phi_R(i)$. Now let us suppose there exists a policy R such that $\phi_R(i) < \phi_{R*}(i)$ for some i. Then there will be an $\varepsilon > 0$ and a subsequence $\{T_v, v = 1, 2, \ldots\}$ of T values such that

$$\frac{S_{R, T_v}(i)}{T_v + 1} < \phi_{R*}(i) - \varepsilon, \qquad v = 1, 2, \ldots.$$

Let us adjoin to K_j for each $j \neq i$ an action $\tilde{a}$ such that $q_{ji}(\tilde{a}) \equiv 1$ with cost $w_{j\tilde{a}} \equiv w$ to be assigned. Let $\tilde{R}$ be defined as above with R_{ji} the policy which takes action $\tilde{a}$ in state j and $T = T_v$ for some v to be assigned. By Theorem 6, Appendix B,

$$\phi_{\tilde{R}}(i) = \lim_{T \to \infty} \frac{S_{\tilde{R}, T}(i)}{T + 1}$$

$$= \frac{S_{R, T_v}(i) + w}{T_v + 2}.$$

Choose w large enough so that

$$\phi_{R*}(i) = \min_{R \in C_D} \phi_R(i);$$

that is, if w is large enough so that any policy $R \in C_D$ using action $\tilde{a}$ is too costly, the policy R^* being originally optimal will again be optimal among all policies in the enlarged class C_D. But also choose v large enough so that

$$\frac{S_{R,T_v}(i) + w}{T_v + 2} < \phi_{R^*}(i) - \frac{\varepsilon}{2},$$

after which we have

$$\phi_R(i) < \phi_{R^*}(i),$$

a contradiction of Theorem 2. This proves the corollary.

Remark: That Theorem 2 and its corollary are true might seem intuitively obvious. This, however, does not stem alone from the Markovian structure we have assumed. For when the state space I is allowed to be countable, Theorem 2 does not hold; moreover, optimal policies may not exist or, when they do exist, they may not be members of C_D or C_S.

First-Passage Problem

We now consider the optimal first-passage problem. Let us first assume that $w_{ia} \geq 0$ for all $a \in K_i$ and $i \in I$. We let $j = 0$ denote a given target state. Without loss of generality, we can take $w_{0a} = 0$ and $q_{00}(a) = 1$ for all $a \in K_0$. For since $j = 0$ is the target state and only costs associated with reaching the target state are relevant, this assumption will not affect the solution to the problem. Then, if $Y_0 = i$ is the initial state and τ denotes the smallest positive integer t such that $Y_n = 0$, then

$$\sigma_R(i) = S_{R,\tau}(i)$$

$$= E_R \sum_{t=0}^{\tau} W_t$$

$$= E_R \sum_{t=0}^{\infty} W_t,$$

where W_t are nonnegative random variables. By setting $w_{0a} = 0$ and

First Passage Problem

$q_{00}(a) = 1$, we are able to remove the random variable τ in the upper limit of the definition of $\sigma_R(i)$.

We now prove:

THEOREM 3. If $\{w_{ia}\}$ are nonnegative, then there exists an $R^* \in C_D$ such that
$$\sigma_{R^*}(i) = \inf_{R \in C} \sigma_R(i), \quad i \in I.$$

Proof: Consider first
$$\Psi_R(i, \alpha) = \sum_{t=0}^{\infty} \alpha^t E_R W_t$$
$$= E_R \sum_{t=0}^{\infty} \alpha^t W_t.$$

The interchange of expectation and summation is justified since $\sum_{t=0}^{T} \alpha^t W_t$ converges uniformly to $\sum_{t=0}^{\infty} \alpha^t W_t$ for $0 \leq \alpha < 1$. By Corollary 1 to Theorem 1 there exists a policy $\tilde{R} \in C_D$ such that $\Psi_{\tilde{R}}(i, \alpha) \leq \Psi_R(i, \alpha)$ for all $i \in I$, $R \in C$, and α near enough to 1. Since for every $R \in C$, $\lim_{\alpha \to 1} \Psi_R(i, \alpha) = \sum_{t=0}^{\infty} E_R W_t$ (the limit may equal ∞), we have from Theorem 4(a) of Appendix B that
$$\sigma_{\tilde{R}}(i) = E_{\tilde{R}} \sum_{t=0}^{\infty} W_t$$
$$= \sum_{t=0}^{\infty} E_{\tilde{R}} W_t,$$

which in turn is
$$\leq \sum_{t=0}^{\infty} E_R W_t$$
$$= E_R \sum_{t=0}^{\infty} W_t$$
$$= \sigma_R(i).$$

The theorem is proved.

In relaxing the condition of nonnegativity on all costs, we need an hypothesis of the sort that $P_R\{Y_0 = 0 \text{ for some } t \geq 1 \mid Y_0 = i\} = 1$ for $i \in I$ for all $R \in C_D$. Such a condition guarantees that $\sigma_R(i)$ is well defined. Without such a condition an infinite path through positive and negative cost states could give rise to an indeterminate cost criterion. We now state and prove Theorem 4, a version of Theorem 3 without the nonnegativity assumption. In Chapter 7 a proof by other methods will also be given for this case.

THEOREM 4. *If $P_R\{Y_t = 0 \text{ for some } t \geq 1 \mid Y_0 = i\} = 1$ for all $i \in I$ and $R \in C_D$, then the conclusion of Theorem 3 holds.*

Proof: Under the hypothesis, the mean first passage m_i^R from state i to 0 is finite for every $i \in I$ and $R \in C_D$ (see Theorem 6 of Appendix A). Now let $w'_{ia} = |w_{ia}|$ and, by Theorem 4(a) of Appendix B,

$$\sigma_R'(i) = E_R \sum_{t=0}^{\infty} W_t'$$

$$= \sum_{t=0}^{\infty} E_R W_t',$$

where $W_t' = w'_{ia}$ if $Y_t = i$, $A_t = a$, $t = 0, 1, \ldots$. By Corollary 1 of Theorem 1 (with maximum replacing minimum) there exists an $R^* \in C_D$ such that

$$\Psi'_{R^*}(i, \alpha) \geq \Psi_R'(i, \alpha), \qquad i \in I, \quad R \in C$$

for all α sufficiently close to 1. However,

$$\sigma_R'(i) = E_R \sum_{t=0}^{\infty} W_t'$$

$$= \sum_{t=0}^{\infty} E_R W_t'$$

$$= \lim_{\alpha \to 1} \sum_{t=0}^{\infty} \alpha^t E_R W_t' \qquad \text{(equation continued)}$$

$$\leq \lim_{\alpha \to 1} \sum_{t=0}^{\infty} \alpha^t E_{R*} W_t'$$

$$= E_{R*} \sum W_t'$$

$$\leq \max_{i,a}\{|w_{ia}| \; m_i^{R*}\}$$

$$< \infty, \quad \text{for every } i \in I \text{ and } R \in C.$$

Thus, since $|W_t| = W_t'$, it follows from Theorem 4(b) of Appendix B that

$$\sigma_R(i) = E_R \sum_{t=0}^{\infty} W_t$$

$$= \sum_{t=0}^{\infty} E_R W_t$$

for every $i \in I$, $R \in C$. Now, again using Corollary 1 to Theorem 1, there exists an $R^* \in C_D$ such that for all $i \in I$, $R \in C$, and α near enough to 1, $\Psi_R(i, \alpha) \geq \Psi_{R*}(i, \alpha)$. Hence, using Theorem 1(a) of Appendix B,

$$\sigma_{R*}(i) = \lim_{\alpha \to 0} \sum_{t=0}^{\infty} \alpha^t E_{R*} W_t$$

$$\leq \lim_{\alpha \to 1} \sum_{t=0}^{\infty} \alpha^t E_R W_t$$

$$= \sigma_R(i), \quad i \neq j, \quad R \in C.$$

This proves the theorem.

First Passage as a Discounted Cost Problem

Consider the special case where $q_{i0}(a) = 1 - \alpha$ for all $a \in K_i$, $i \in I - \{0\}$, where again $q_{00}(a) \equiv 1$, $w_{0a} \equiv 0$. Then, for any $R \in C$,

$$P_R\{\tau = t \mid Y_0 = i\} = \alpha^{t-1}(1 - \alpha), \quad t = 1, 2, \ldots, \quad i \in I - \{0\}.$$

But

$$\sigma_R(i) = \sum_{T=1}^{\infty} E_R \left(\sum_{t=0}^{T-1} W_t \,|\, \tau = T, \quad Y_0 = i \right) \alpha^{T-1}(1-\alpha)$$

$$= \sum_{T=1}^{\infty} \sum_{t=0}^{T-1} E_R(W_t \,|\, \tau = T, \quad Y_0 = i) \alpha^{T-1}(1-\alpha)$$

$$= \sum_{t=0}^{\infty} \sum_{T=t+1}^{\infty} E_R(W_t \,|\, \tau = T, \quad Y_0 = i) \alpha^{T-1}(1-\alpha)$$

$$= \sum_{t=0}^{\infty} E_R\{W_t \,|\, \tau > t, \quad Y_0 = i\} P\{\tau > t\}$$

$$= \sum_{t=0}^{\infty} E_R\{W_t \,|\, \tau > t, \quad Y_0 = i\} \alpha^t, \qquad i \in I - \{0\}.$$

However,

$$E\{W_t \,|\, \tau > t, \quad Y_0 = i\}$$
$$= \sum_{j \neq 0} \sum_a w_{ja} P_R\{Y_t = j, \quad A_t = a \,|\, Y_0 = i\} \,/\, \alpha^t, \qquad i \in I - \{0\}.$$

Thus, $\sigma_R(i)$, $i \in I - \{0\}$, is equivalent to $\Psi_R(i, \alpha)$, $i \in I - \{0\}$, for a Markovian decision process based on the state space $I' = I - \{0\}$, with laws of motion $q'_{ij}(a) = q'_{ij}(a)/\alpha$, $a \in K_i$, $i, j \in I'$, and costs $w'_{ia} = w_{ia}$, $a \in K_i$, $i \in I'$. Or, let $\Psi_R(i, \alpha)$ be a discounted cost criterion over a state space I, with laws of motion $\{q_{ij}(a)\}$. Define a fictitious state "0" and adjoin it to I. Define laws of motion by $q'_{i0}(a) = 1 - \alpha$, $i \in I$, $q'_{ij}(a) = \alpha q_{ij}(a)$, $a \in K_i$, $i, j \in I$, with $q_{00}(a) \equiv 1$. Then, $\Psi_R(i, \alpha)$ is equivalent to $\sigma_R(i)$, $i \in I$, where $\sigma_R(i)$ is the expected cost of a first passage to the state "0" using the laws of motion $\{q'_{ij}(a)\}$.

Bibliographical Remarks

Using essentially the same method, Derman [17] proved Theorem 1 for the case of a denumerable state space I. The idea of using Tychonov's theorem in this connection is due to Karlin [35]. Another proof of

Theorem 1 holding for a denumerable state space is given by Blackwell [7].

Corollary 1 to Theorem 1 is due to Blackwell [6].

The proof of Theorem 2 employs techniques used in Derman [15] and Gillette [33]. Gillette [33] employs an incorrect (see Liggett and Lippman [42]) extension of a theorem by Hardy and Littlewood while working with $\phi_R(i)$ defined by $\lim_{T\to\infty} \inf S_{R,T}(i)/T$. The proof of Corollary 1 to Theorem 2 given here is a modification of the one given by Derman [18].

For more on the remark following Theorem 2 see Derman [19] and Fisher and Ross [32].

Theorem 3 was proved by Derman [15].

In connection with Problem 2 following, Veinott [53] provides an alternative proof. His proof shows that the limit converges geometrically. This result in Problem 2 is used in the proof of Theorem 1 of Chapter 5.

Problems

(1) Is $\phi_R(i)$ always a continuous function of R?

(2) Suppose $q_{00}(a) = 1$ for every $a \in K_0$ and that $P_R\{Y_t = 0 \text{ for some } t > 0 \mid Y_0 = i\} > 0$, $i \in I$, $R \in C_D$. Show that

$$\limsup_{t\to\infty, R\in C} P_R\{Y_t \neq 0 \mid Y_0 = i\} = 0, \quad i \in I.$$

Solution: By Theorem 4 and Theorem 6 of Appendix A, on maximizing the expected first-passage cost with $w_{ia} = 1$, $a \in K_i$, $i \neq 0$, there exists an $R^* \in C_D$ such that

$$m_i^{R^*} = \max_{R \in C} m_i^R < \infty,$$

where m_i^R denotes the mean first-passage time to 0 from

state i. But by Chebychev's inequality, $(P\{X > c\} < EX/c$ if X is a nonnegative random variable) for every $R \in C$, $i \in I$,

$$P_R\{Y_t \neq 0 \mid Y_0 = i\} = P_R\{(\text{first-passage time to } 0) > t \mid Y_0 = i\}$$

$$\leq \frac{m_i^R}{t}$$

$$\leq \frac{m^{R*}}{t}.$$

Thus, for $i \in I$,

$$\sup_{R \in C} P_R\{Y_t \neq 0 \mid Y_0 = i\} \leq \frac{m_i^{R*}}{t}$$

and

$$\lim_{t \to \infty} \sup_{R \in C} P_R\{Y_t \neq 0 \mid Y_0 = i\} = 0, \qquad i \in I.$$

4

Computational Methods for the Discounted Cost Problem

Introduction and Summary

In the previous chapter we showed that solutions exist to the problems of minimizing the expected discounted cost and the expected average cost criterion as well as to the problem of minimizing the mean first-passage cost. In this chapter and Chapters 5 and 6, we present methods by which the optimal policies actually can be obtained.

Although it was shown that optimal policies for each problem exist in C_D, the problem of finding them is nontrivial. In spite of the fact that, given $R \in C_D$, the cost criteria Ψ_R, ϕ_R, and σ_R can be evaluated, the number of policies in C_D may be astronomically large. For example if I contains N states and $K_i = 2$, $i \in I$, then C_D contains 2^N different policies. For small values of N the method of simple enumeration is feasible; however, for N moderately large, complete enumeration is virtually impossible. Nevertheless, modern computational methods have overcome problems of this sort. For example, in solving the linear programming problem, the space of possible solutions that would have to be searched if the problem were to be solved by enumeration would usually be a large finite number; however, computational methods (the simplex method is one) have been developed which select an optimal solution without the need for complete enumeration. We shall show here that comparable methods exist for obtaining optimal policies. In fact, linear programming computational procedures, among others, can be employed.

In this chapter we discuss the problem of minimizing the expected discounted cost criterion Ψ_R. We present three approaches to obtaining the optimal policy: the method of successive approximations, policy improvement, and linear programming. The first is the classical method used in solving differential and integral equations. In itself, it does not provide a method for obtaining a solution in a finite number of iterations; however, slightly modified, it can. More significantly, this method has some uses in determining structure of optimal policies. Both the policy improvement and linear programming methods are finite algorithms and are feasible, provided the size of I is not too large.

Method of Successive Approximations

Let $\{v_0(i), i \in I\}$ denote an arbitrary set of values. Define for $n = 0, 1, \ldots,$

$$v_{n+1}(i) = \min_a \left\{ w_{ia} + \alpha \sum_j q_{ij}(a) v_n(j) \right\}, \qquad i \in I, \tag{1}$$

Method of Successive Approximations

where $0 \leq \alpha 1$ is fixed. We have:

THEOREM 1. If $\{v_n(i), i \in I, n = 1, \ldots\}$ is defined by the transformation (1) with $\{v_0(i), i \in I\}$ arbitrary, then $\lim_{n \to \infty} v_n(i) = \Psi_{R_0}(i, \alpha), i \in I$, independent of $\{v_0(i), i \in I\}$, where R_0 is a policy that minimizes $\Psi_R(i, \alpha), i \in I$.

Proof: Let $\{v_0(i), i \in I\}$ be arbitrary and $v_0'(i) = \Psi_{R_0}(i, \alpha), i \in I$. We first show that

$$\max_{i \in I} |v_{n+1}(i) - v'_{n+1}(i)| \leq \alpha \max_{i \in I} |v_n(i) - v_n'(i)|, \qquad n = 0, 1, \ldots.$$

Let a_i', $i \in I$, be the action which minimizes the right-hand side of Eq. (1), where in (1) $v_n(i)$ is the nth iterate of $v_0'(i)$. Then

$$v_{n+1}(i) - v'_{n+1}(i) \leq w_{ia_i'} + \alpha \sum_j q_{ij}(a_i') v_n(j) - w_{ia_i'} - \alpha \sum_j q_{ij}(a_i') v_n'(j)$$

$$\leq \alpha \sum_j q_{ij}(a_i') \max_j |v_n(j) - v_n'(j)|$$

$$= \alpha \max_j |v_n(j) - v_n'(j)|, \qquad i \in I.$$

Similarly, on letting a_i, $i \in I$, be the action which minimizes the right-hand side of Eq. (1), where in (1) $v_n(i)$ is the nth iterate of $v_0(i)$, we obtain

$$v'_{n+1}(i) - v_{n+1}(i) \leq \alpha \max_j |v_n(j) - v_n'(j)|, \qquad i \in I.$$

Hence, we have shown the inequality. Now, by iteration, we obtain that

$$|v'_{n+1}(i) - v_{n+1}(i)| \leq \alpha^n \max_j |v_0(j) - v_0'(j)|, \qquad i \in I, \quad n = 0, 1, \ldots.$$

Thus, $\lim_{n \to \infty} (v_n'(i) - v_n(i)) = 0, i \in I$. However, from Theorem 1 of Chapter 3 (without using uniqueness) we have that $v_n'(i) = \Psi_{R_0}(i, \alpha)$, $i \in I, n = 0, 1, \ldots$. Hence, $\lim_{n \to \infty} v_n(i) = \Psi_{R_0}(i, \alpha), i \in I$. Since $v_0(i), i \in I$, was arbitrarily selected, the theorem is proven.

The proof of the uniqueness part of Theorem 1 of Chapter 3 was postponed. Essentially, we now have shown this and summarize it in

COROLLARY 1. Equation (2) of Chapter 3* has one and only one solution, namely, $\Psi_{R_0}(i, \alpha)$, $i \in I$, where R_0 is a policy that minimizes $\Psi_R(i, \alpha)$, $i \in I$.

Proof: If $v_0''(i)$, $i \in I$, is a second solution, then $v_n''(i) = v_0''(i)$, $i \in I$, $n = 1, \ldots,$ and thus, $\lim_{n \to \infty} v_n''(i) = v_0''(i)$. However, from Theorem 1, $\lim_{n \to \infty} v_n''(i)$ has been evaluated to be $\Psi_{R_0}(i, \alpha)$.

The method of successive approximations based on Theorem 1 consists of selecting an initial arbitrary function and transforming it successively according to the transformation defined by (1). The limiting function will satisfy Eq. (3.2) and the optimal policy is obtained by taking that action in state i, $i \in I$, which minimizes the right-hand side of (3.2). In practice, the limiting function will be approximated only. An approximation to the optimal policy is obtained by treating $\Psi_{R_0}(i, \alpha)$ and its approximation as if they were equal. Actually, if the approximation is close enough to $\Psi_{R_0}(i, \alpha)$, which will be the case for large n, the exact optimal policy will be obtained. However, within the procedure, no formal stopping method is given. One might modify the method by occasionally evaluating for various values of n, $\Psi_{R_n}(i, \alpha)$, $i \in I$, from Eq. (2) following, where $R_n \in C_D$ is the policy defined by taking that action in state i, $i \in I$, which minimizes the right-hand side of (1). If $\{\Psi_{R_n}(i, \alpha), i \in I\}$ satisfies (3.2) then R_n is optimal. See also Problems 2, 3, and 4 at the end of this chapter.

In general, the functions $\{v_n(i), i \in I\}$, $n = 0, 1, \ldots$ have no tangible interpretation. However, if $v_0(i) = \min\{w_{ia}\}$, $i \in I$, then using methods of Chapter 2, we have

$$v_n(i) = \inf_{R \in C} E_R\left(\sum_{t=0}^{n} \alpha^t W_t \mid Y_0 = i\right), \qquad n = 0, 1, \ldots,$$

* Hereafter this equation will be referred to as (3.2).

Policy Improvement Procedure

the optimal expected discounted cost criterion over the periods $0, 1, \ldots, n$. Otherwise, if $v_0(i)$ is interpreted as the terminal cost of being in state i at time n if the process is terminated at time n, then $v_n(i)$ is the minimal expected discounted cost plus terminal cost over periods $0, 1, \ldots, n$ given that the initial state is i.

In practice, the method of successive approximations may be used when an approximation or guess to an expected discounted cost criterion corresponding to a heuristic policy (one arrived at by respected intuition) is available. Then several iterations will hopefully improve it. In any case, use of the method of successive approximations never necessitates the computation of an exact discounted cost criterion; thus, one is spared the work of solving the system of simultaneous equations (2) following. This latter feature makes the method an attractive computational procedure, particularly if the computations are done by hand or with a desk calculator.

Perhaps the method of successive approximations is of greatest value in a more theoretical context. That is, certain mathematical properties can be ascertained. For example, suppose I is the set of integers $0, 1, \ldots, L$ and the laws of motion are such that $v_1(i)$ is a nondecreasing function of i whenever $v_0(i)$ is a nondecreasing function. Then it follows that $\Psi_{R_0}(i, \alpha)$ is a nondecreasing function. From this property the structure of an optimal policy may sometimes be deduced. (See Chapter 9, Section 1.)

Policy Improvement Procedure

This is an iterative procedure that improves on each iteration and terminates after a finite number of iterations with an optimal policy.

Let $R_1 \in C_D$ be arbitrary. Then $\{\Psi_{R_1}(i, \alpha), i \in I\}$ satisfies, uniquely, the equations

$$\Psi_{R_1}(i, \alpha) = w_{iR_1} + \alpha \sum_j q_{ij}(R_1)\Psi_{R_1}(j, \alpha), \qquad i \in I. \tag{2}$$

Uniqueness follows from the fact (using Theorem 3 of Appendix A) that the matrix of system (2) is $I - \alpha Q$ (I is the identity matrix and

$Q = \{q_{ij}(R_1)\}$) which has an inverse $\{I - \alpha Q\}^{-1} = \sum_{n=0}^{\infty} \alpha^n Q^n$. Let E_i denote the set of actions i for which $w_{ia} + \alpha \sum_j q_{ij}(a)\Psi_{R_1}(j, \alpha)$ is strictly less than the right-hand side of (2). Define $R_2 \in C_D$ as follows: For one or more states i for which E_i is nonempty, prescribe action a in E_i. For all other states, take the action prescribed by R_1. We refer to the derivation of R_2 from R_1 as a *policy improvement iteration*. The fact that the iteration is an improvement is established in the following.

THEOREM 2. If E_i is nonempty for at least one state i, then $\Psi_{R_2}(i, \alpha) \leq \Psi_{R_1}(i, \alpha)$, $i \in I$, with strict inequality holding at every i for which $R_2 \neq R_1$.

Proof: By definition of the policy improvement iteration,

$$\Psi_{R_1}(i, \alpha) \geq w_{iR_2} + \alpha \sum_j q_{ij}(R_2)\Psi_{R_1}(j, \alpha), \qquad i \in I, \qquad (3)$$

with strict inequality holding at each i for which $R_2 \neq R_1$. Let $\{q_{ij}^{(t)}(R_2)\}$ ($t = 0, 1, \ldots$) denote the t-step transition probabilities under R_2. Then from (3), on premultiplying by $\alpha^t q_{ji}^{(t)}(R_2)$ and summing over i, we can write for $t = 0, 1, \ldots$,

$$\alpha^t \sum_i q_{ji}^{(t)}(R_2)\Psi_{R_1}(i, \alpha)$$
$$\geq \alpha^t \sum_i q_{ji}^{(t)}(R_2)w_{iR_2} + \alpha^{t+1} \sum_l q_{jl}^{(t+1)}(R_2)\Psi_{R_1}(l, \alpha), \qquad j \in I. \qquad (4)$$

For $t = 0$, Eqs. (3) and (4) are identical. On summing (4) over $t = 0, 1, \ldots$, we obtain, since $\Psi_{R_2}(j, \alpha) = \sum_{t=0}^{\infty} \alpha^t \sum_i q_{ji}^{(t)}(R_2)w_{iR_2}$,

$$\sum_{t=0}^{\infty} \alpha^t \sum_i q_{ji}^{(t)}(R_2)\Psi_{R_1}(i, \alpha)$$
$$\geq \Psi_{R_2}(j, \alpha) + \sum_{t=1}^{\infty} \alpha^t \sum_l q_{jl}^{(t)}(R_2)\Psi_{R_0}(l, \alpha), \qquad j \in I,$$

Linear Programming

with strict inequality holding, because of terms at $t = 0$, at each j for which $R_2 \neq R_1$. On subtracting the second term on the right-hand side from the left term, and because the two differ only when $j = i$ (since $q_{ij}^{(0)} = \delta_{ij}$), we have $\Psi_{R_1}(j, \alpha) \geqq \Psi_{R_2}(j, \alpha), j \in I$, with strict inequality holding for each j for which $R_1 \neq R_2$. Thus the theorem is proved.

We refer to a sequence of policy improvement iterations as the *policy improvement procedure*. We can state

COROLLARY 1. The policy improvement procedure terminates, after a finite number of iterations, at an optimal policy.

Proof: C_D contains only a finite number of policies. Since each iteration is accompanied by a strict improvement, no repetitions will occur. Thus, at some point no improvements will be possible, at which time (3.2) will hold and the corollary is proved.

In summary, the policy improvement procedure provides a monotone (always improving) convergent sequence of policies and attains in a finite number of iterations the optimal policy. Its drawback is that the discounted cost function for each policy R in the sequence must be calculated. This involves solving the linear system (2).

Linear Programming

Both the methods of successive approximation and policy improvement may be regarded as methods of dynamic programming. Thus, it is somewhat surprising that the method of linear programming can also be brought to bear. For this, consider the linear programming problem:

Maximize
$$\sum_j \beta_j v_j$$

subject to

$$v_i \leq w_{ia} + \alpha \sum_j q_{ij}(a)v_j, \qquad a \in K_i, \quad i \in I,$$

where $\beta_j > 0, j \in I$, and $\sum_j \beta_j = 1$ are given numbers. The dual linear programming problem is:

Minimize

$$\sum_i \sum_a x_{ia} w_{ia}$$

subject to

$$x_{ia} \geq 0, \qquad a \in K_i, \quad i \in I,$$

and

$$\sum_i \sum_a x_{ia}(\delta_{ij} - \alpha q_{ij}(a)) = \beta_j, \qquad j \in I,$$

where $\delta_{ij} = 0$ if $i \neq j$, and 1 if $i = j$.

We first discuss the dual problem.

THEOREM 3. Let $R \in C_S$ be defined by $\{D_{ia}\}$: Then

$$\tilde{x}_{ia} = \sum_l \beta_l \sum_{t=0}^{\infty} \alpha^t P_R\{Y_t = i \mid Y_0 = l\} D_{ia}, \qquad a \in K_i, \quad i \in I,$$

is a feasible solution to the dual problem. On the other hand, if $\{x_{ia}\}$ is any feasible solution to the dual problem, then $\{D_{ia}\} = \{x_{ia}/\sum_{a'} x_{ia'}\}$ defines a policy $R \in C_S$ and $x_{ia} = \tilde{x}_{ia}, a \in K_i, i \in I$. That is, $\{D_{ia}\} = \{x_{ia}/\sum_{a'} x_{ia'}\}$ is a one-to-one mapping of the feasible solutions to the dual problem onto C_S.

Proof: It can be readily verified that $\{\tilde{x}_{ia}\}$ satisfies the feasibility constraints. Now let $\{x_{ia}\}$ be any feasible solution to the dual problem.

Linear Programming

The feasibility equations can be written

$$\sum_a x_{ja} - \alpha \sum_i \left(\sum_a x_{ia} \right) \sum_{a'} q_{ij}(a') D_{ia'} = \beta_j, \quad j \in I,$$

from which it follows that $\sum_a x_{ia} > 0, i \in I$, since it is assumed that $\beta_j > 0, j \in I$. Thus $\{D_{ia}\} = \{x_{ia}/\sum_a x_{ia}\}$ is well defined. However, treating $\sum_a x_{ia}, i \in I$, as variables in the above representation of the feasibility equations, it follows from Theorem 3, Appendix A that

$$\sum_a x_{ia} = \sum_a \tilde{x}_{ia}, \quad i \in I.$$

But

$$\tilde{x}_{ia} = \sum_{a'} \tilde{x}_{ia'} D_{ia}$$

$$= \sum_{a'} x_{ia'} D_{ia}$$

$$= x_{ia}, \quad a \in K_i, \quad i \in I,$$

which completes the proof of the theorem.

COROLLARY 1. An optimal policy $R_0 \in C_S$ is obtained by solving the dual problem and setting $D_{ia}^{R_0} = x_{ia}/\sum_a x_{ia}, a \in K_i, i \in I$, where $\{x_{ia}\}$ is an optimal solution to the dual problem.

Proof: Since the objective function of the dual problem is in fact $\sum_{l \in I} \beta_l \Psi_R(l, \alpha)$, this expression is minimized. However, since $\beta_l > 0$ and a single $R \in C_D$ minimizes $\Psi_R(l, \alpha)$ for every $l \in I$, it follows that $\Psi_R(l, \alpha)$ is minimized for each $l \in I$.

COROLLARY 2. If the simplex method is used to solve the dual problem, an optimal policy $R_0 \in C_D$ is obtained.

Proof: The simplex method obtains only extreme point solutions. It follows that $\sum_a x_{ia} > 0$, $i \in I$, and from Theorem 3 of Appendix C, $x_{ia} > 0$ for exactly one $a \in K_i$ for each $i \in I$. This then implies $D_{ia}^{R_0} = 1$ or 0, $i \in I$.

COROLLARY 3. For every optimal solution to the primal problem, (3.2) must hold.

Proof: It follows from the complementary slackness property of primal and dual linear programming problems (Theorem 5, Appendix C) that if $\{v_j, j \in I\}$ is optimal for the primal problem, then

$$v_i = w_{ia} + \alpha \sum_j q_{ij}(a) v_j$$

for those values of i and a where $x_{ia} > 0$. However, we have seen in the proof of Theorem 3 that for each $i \in I$, $x_{ia} > 0$ for some a. Therefore, if we consider the constraints of the primal problem, (3.2) must hold.

We can now prove

THEOREM 4. If $\{v_j^0\}$ is an optimal solution to the primal problem, then $\{v_j^0\}$ satisfies (3.2) and consequently $v_j^0 = \Psi_{R_0}(j, \alpha), j \in I$, where $R_0 \in C_D$ is an optimal policy.

Proof: From Corollary 3 we have that (3.2) must be satisfied by an optimal solution to the primal problem. Since by corollary 1 to Theorem 1, Eq. (3.2) has a unique solution $\{\Psi_{R_0}(j, \alpha), j \in I\}$, the equality must follow.

COROLLARY 1. An optimal policy $R_0 \in C_D$ can be obtained from the optimal solution to the primal problem by letting R_0 be the policy that takes action $a = a_i$ at state i which achieves equality in the constraints of the primal linear programming problem. If more than one action a achieves equality at any state then either action may be taken.

Proof: Once a solution to (3.2) is obtained by any means, an optimal policy is prescribed according to Theorem 1 of Chapter 3.

Theorems 3 and 4 and their corollaries provide the linear programming machinery for obtaining optimal policies. We point out that the variables $\{x_{ia}\}$ of the dual problem have policy and expected frequency interpretations for *every* feasible solution to the problem.[1] For x_{ia} is, in a discounted sense, an average probability of being in state i and making decision a when $P\{Y_0 = l\} = \beta_l, l \in I$, and policy $R \in C_S$ is used, where R is given by $D_{ia} = x_{ia}/\sum_a x_{ia}$. Thus, in a sense, the simplex algorithm for solving the dual problem, which is a procedure that has the property that successive iterations provide improving solutions, is a special type of policy improvement method. On the other hand, for the primal problem, it is the equalities (3.2) obtained only in an optimal solution that yield an interpretation with respect to policies. Of course, the optimal values have their interpretation in terms of being the optimal discounted costs $\Psi_R(i, \alpha), i \in I$.

In the proof of Theorem 3, the assumption that $\beta_l > 0, l \in I$, is used in showing that $\sum_a x_{ia} > 0, i \in I$. If we allow a subset S of states such that $\beta_l = 0, l \in S$, then it is possible that $\sum_a x_{ia} = 0$ for some $i \in S$. In this case, suppose we define $R = \{D_{ia}\}$ by setting $D_{ia} = x_{ia}/\sum_a x_{ia}$ if $\sum_a x_{ia} > 0$ and choosing D_{ia} arbitrarily if $\sum_a x_{ia} = 0$. Then, for every i such that $\sum_a x_{ia} = 0$, we can assert that $P_R\{Y_t = l \mid Y_0 = l\} = 0$, $t = 0, 1, \ldots$ for every l such that $\beta_l > 0$. Moreover, R is optimal with respect to minimizing $\Psi_R(l, \alpha)$ for every l for which $\beta_l > 0$. However, R may not be optimal with respect to minimizing $\Psi_R(i, \alpha)$ for every i such that $\sum_a x_{ia} = 0$.

---

[1] For this reason we should have perhaps referred to the problem involving the variables $\{x_{ia}\}$ as the primal linear programming problem and to the other as dual. However, since the problem involving the variables $\{v_i\}$ arises first we have called it primal.

Computational Examples

Suppose, as in Chapter 2, that $I = \{0, 1\}$, $K_i = 2$, $i = 0, 1$, where

$$\begin{Bmatrix} w_{01} & w_{02} \\ w_{11} & w_{12} \end{Bmatrix} = \begin{Bmatrix} 1 & 0 \\ 2 & 2 \end{Bmatrix}$$

and

$$\begin{Bmatrix} (q_{00}(1), q_{00}(2)) & (q_{01}(1), q_{01}(2)) \\ (q_{10}(1), q_{10}(2)) & (q_{11}(1), q_{11}(2)) \end{Bmatrix} = \begin{Bmatrix} (\tfrac{1}{2}, \tfrac{1}{4}) & (\tfrac{1}{2}, \tfrac{3}{4}) \\ (\tfrac{2}{3}, \tfrac{1}{3}) & (\tfrac{1}{3}, \tfrac{2}{3}) \end{Bmatrix}.$$

We take $\alpha = \tfrac{1}{2}$.

Let us first employ the method of successive approximations in order to obtain an optimal policy. Let $v_0(0) = v_0(1) = 0$. Then using (1),

$$v_1(0) = \min_a \left\{ w_{0a} + \alpha \sum_j q_{0j}(a) v_0(j) \right\}$$
$$= \min\{1, 0\}$$
$$= 0,$$

and, similarly,

$$v_1(0) = \min\{2, 2\}$$
$$= 2.$$

Then,

$$v_2(0) = \min\left\{1 + \tfrac{1}{2} \cdot \tfrac{1}{2} \cdot 2,\ 0 + \tfrac{1}{2} \cdot \tfrac{3}{4} \cdot 2\right\}$$
$$= \min\left\{\tfrac{3}{2}, \tfrac{3}{4}\right\}$$
$$= \tfrac{3}{4}$$

Computational Examples

and

$$v_2(1) = \min\left\{2 + \frac{1}{2}\cdot\frac{1}{3}\cdot 2, 2 + \frac{1}{2}\cdot\frac{2}{3}\cdot 2\right\}$$

$$= \min\left\{\frac{7}{3}, \frac{8}{3}\right\}$$

$$= \frac{7}{3}.$$

Iteration once again yields

$$v_3(0) = \min\left\{1 + \frac{1}{2}\left(\frac{1}{2}\cdot\frac{3}{4} + \frac{1}{2}\cdot\frac{7}{3}\right), \frac{1}{2}\left(\frac{1}{4}\cdot\frac{3}{4} + \frac{3}{4}\cdot\frac{7}{3}\right)\right\}$$

$$= \min\left\{\frac{37}{24}, \frac{93}{96}\right\}$$

$$= \frac{31}{32}$$

and

$$v_3(1) = \min\left\{2 + \frac{1}{2}\left(\frac{2}{3}\cdot\frac{3}{4} + \frac{1}{3}\cdot\frac{7}{3}\right), 2 + \frac{1}{2}\left(\frac{1}{3}\cdot\frac{3}{4} + \frac{2}{3}\cdot\frac{7}{3}\right)\right\}$$

$$= 2 + \frac{1}{2}\left(\frac{2}{3}\cdot\frac{3}{4} + \frac{1}{3}\cdot\frac{7}{3}\right)$$

$$= \frac{95}{36}.$$

The policy R^* approximating the optimal policy is the one that takes action $a = 2$ at state 0 since

$$1 + \frac{1}{2}\left(\frac{1}{2}\cdot\frac{31}{32} + \frac{1}{2}\cdot\frac{95}{36}\right) > \frac{1}{2}\left(\frac{1}{4}\cdot\frac{31}{32} + \frac{3}{4}\cdot\frac{95}{36}\right),$$

and takes action $a = 1$ at state 1 since

$$2 + \frac{1}{2}\left(\frac{2}{3}\cdot\frac{31}{32} + \frac{1}{3}\cdot\frac{95}{36}\right) < 2 + \frac{1}{2}\left(\frac{1}{3}\cdot\frac{31}{32} + \frac{2}{3}\cdot\frac{95}{36}\right).$$

Let us check whether this policy is, in fact, optimal. We have that
$$\Psi_{R*}(0) = \tfrac{1}{2} \cdot (\tfrac{1}{4}\Psi_{R*}(0) + \tfrac{3}{4}\Psi_{R*}(1))$$
and
$$\Psi_{R*}(1) = 2 + \tfrac{1}{2}(\tfrac{2}{3}\Psi_{R*}(0) + \tfrac{1}{3}\Psi_{R*}(1)).$$
On solving, we obtain
$$\Psi_{R*}(0) = \frac{36}{29}, \qquad \Psi_{R*}(1) = \frac{84}{29}.$$
We can now check whether
$$1 + \frac{1}{2}\left(\frac{1}{2} \cdot \frac{36}{29} + \frac{1}{2} \cdot \frac{84}{29}\right) > \frac{1}{2}\left(\frac{1}{4} \cdot \frac{36}{29} + \frac{3}{4} \cdot \frac{84}{29}\right)$$
and
$$2 + \frac{1}{2}\left(\frac{2}{3} \cdot \frac{36}{29} + \frac{1}{3} \cdot \frac{84}{29}\right) < 2 + \frac{1}{2}\left(\frac{1}{3} \cdot \frac{36}{29} + \frac{2}{3} \cdot \frac{84}{29}\right).$$

The inequalities hold so that R^* is, in fact, optimal.

Let us now obtain an optimal policy using the policy improvement procedure. Let R_1 be the policy that takes action $a = 1$ at state 0 and action $a = 1$ at state $a = 1$. Then
$$\Psi_{R_1}(0) = 1 + \tfrac{1}{2}(\tfrac{1}{2}\Psi_{R_1}(0) + \tfrac{1}{2}\Psi_{R_1}(1)),$$
$$\Psi_{R_1}(1) = 2 + \tfrac{1}{2}(\tfrac{2}{3}\Psi_{R_1}(0) + \tfrac{1}{3}\Psi_{R_1}(1)),$$
from which we see that
$$\Psi_{R_1}(0) = \frac{32}{13}, \qquad \Psi_{R_1}(1) = \frac{44}{13}.$$
At state 0,
$$0 + \frac{1}{2}\left(\frac{1}{4} \cdot \frac{32}{13} + \frac{3}{4} \cdot \frac{44}{13}\right) < \frac{32}{13};$$
hence, the policy R_2, action $a = 2$ at state 0, and action $a = 1$ at state 1, is better. Since we already know R_2 is optimal, no further policy improvement iterations will be possible.

The primal linear programming problem, letting $\beta_0 = \beta_1 = \frac{1}{2}$, is:

To maximize
$$\tfrac{1}{2}(v_0 + v_1)$$
subject to
$$\tfrac{3}{4}v_0 - \tfrac{1}{4}v_1 \leq 1,$$
$$\tfrac{7}{8}v_0 - \tfrac{3}{8}v_1 \leq 0,$$
$$-\tfrac{1}{3}v_0 + \tfrac{5}{6}v_1 \leq 2,$$
$$-\tfrac{1}{6}v_0 + \tfrac{2}{3}v_1 \leq 2.$$

The dual problem is to minimize
$$x_{01} + 2x_{11} + 2x_{12}$$
subject to
$$x_{01} \geq 0, \quad x_{02} \geq 0, \quad x_{11} \geq 0, \quad x_{12} \geq 0$$
and
$$\tfrac{3}{4}x_{01} + \tfrac{7}{8}x_{02} - \tfrac{1}{3}x_{11} - \tfrac{1}{6}x_{12} = \tfrac{1}{2},$$
$$-\tfrac{1}{4}x_{01} - \tfrac{3}{8}x_{02} + \tfrac{5}{6}x_{11} + \tfrac{2}{3}x_{12} = \tfrac{1}{2}.$$

We leave it to the reader to numerically solve each of the linear programming problems and determine the optimal policy from each solution.

Bibliographical Remarks

The proof of Theorem 1 essentially involves establishing that the transformation defined by (1) is a contraction. Since we know a fixed point of (1) already exists; namely $\{\Psi_{R_0}(i), i \in I\}$, the remainder of the proof is somewhat simpler than the classic proof of the Picard–Banach fixed point theorem. Maitra's proof [45] for the denumerable state case motivated our approach. However, the theorem for the finite case should be credited to Shapley [48] who also used the contraction method.

That the policy improvement procedures are associated with dynamic programming can be seen in the writings of Bellman (see, for example, [3]). The explicit procedure for the Markovian decision process with discounted cost criterion appears in Howard [34]. See also Blackwell [6].

That linear programming can be used for the discounted cost criterion is due to D'Epenoux [14].

Problems

(1) Using the data provided for Problem 1 of Chapter 2, find the minimal expected discounted cost policy using each of the computational methods.

(2) In the method of successive approximations with $v_0(i) = 0$, $i \in I$, show that for each $i \in I$,

$$|v_{n+1}(i) - \Psi_{R_0}(i, \alpha)| \leq \frac{\alpha^n}{1 - \alpha} \max_{i, a} \{|w_{ia}|\},$$

where R_0 is optimal.

(3) Assume $\min_{\substack{i, a, a' \\ a' \neq a}} \{|w_{ia} - w_{ia'}|\} > 0$. How large must n be in order that the method of successive approximations yields an optimal policy?

(4) If in the method of successive approximations, $v_1(i) \geq v_0(i)$, $i \in I$, show that $v_{n+1}(i) \geq v_n(i)$, $i \in I$; that is, that $\{v_n(i), n = 0, 1, \ldots\}$ converges monotonically to $\Psi_{R_0}(i, \alpha)$ from below.

(5) By direct argument, that is, without resorting to the dual problem, prove that the optimal solution $\{v_i, i \in I\}$ to the primal problem must satisfy

$$v_j = \min_a \left\{ w_{ia} + \sum_j q_{ij}(a) v_j \right\}, \quad i \in I.$$

Problems

(6) Define the transformation T_R of a vector $v = \{v(i), i \in I\}$ by

$$(T_R v)(i) = w_{iR} + \alpha \sum_j q_{ij}(R)v(j), \qquad i \in I.$$

Show that if $v'(i) \geqq v(i)$, $i \in I$, then $(T_R v')(i) \geqq (T_R v)(i)$, $i \in I$.

(7) Prove the assertions in the last paragraph of the linear programming section.

5

Computational Procedures for the Optimal First-Passage Problem

Introduction

In solving the optimal policy for the expected discounted cost criterion, we saw that the methods of successive approximation, policy improvement, and linear programming can all be used. The same can be said for obtaining optimal solutions to the optimal mean first-passage problem.

We assume that state $j = 0$ is the target state and $P_R\{Y_t = 0$ for some $\{t > 0 \mid Y_0 = i\} = 1$, $i \in I$, and $q_{00}(R) = 1$ for all $R \in C_D$. Recall from Theorem 4, Chapter 3 that some policy $R \in C_D$ is optimal and thus we need only consider the rules in C_D. However, as in the previous chapter, it will be convenient, in the linear programming formulation, to consider the class C_S of policies.

Method of Successive Approximations

We first discuss the method of successive approximations in the present context. Let $\{v_0(i), i \in I - \{0\}\}$ be arbitrary, and define

$$v_{n+1}(i) = \min_a \left\{ w_{ia} + \sum_{j \neq 0} q_{ij}(a) v_n(j) \right\}, \quad i \in I - \{0\}. \qquad (1)$$

We shall prove:

THEOREM 1. If $\{v_n(i), i \in I - \{0\}, n = 0, 1, \ldots\}$ are defined by transformation (1), then $\lim_{n \to \infty} v_n(i) = \sigma_{R_0}(i)$, $i \in I - \{0\}$, independent of $\{v_0(i), i \in I - \{0\}\}$, where $R_0 \in C_D$ minimizes $\sigma_R(i)$, $i \in I - \{0\}$.

Proof: Let $\{v_0(i), i \in I - \{0\}\}$ be arbitrary and $v_0'(i) = \sigma_{R_0}(i)$, $i \in I - \{0\}$. Since R_0 is optimal and is a member of C_D we have that $v_n'(i)$, the nth iterate of $v_0'(i)$ in (1), is equal to $v_0'(i)$ for $n = 1, 2, \ldots$. Let a_i, $i \in I - \{0\}$ denote the actions minimizing the right-hand side of (1) where $v_n(i)$ is the nth iterate of $v_0(i)$ in (1). On subtraction, we have for $n = 0, 1, \ldots$,

$$v'_{n+1}(i) - v_{n+1}(i) \leq \sum_{j \neq 0} q_{ij}(a_i) |v_n(j) - v_n'(j)|, \quad i \in I - \{0\}.$$

Similarly, if a_i', $i \in I - \{0\}$, denote the actions minimizing the right-hand side of (1), where $v_n(i)$ of (1) is the nth iterate of $v_0'(i)$. Then for $n = 0, 1, \ldots$,

$$v_{n+1}(i) - v'_{n+1}(i) \leq \sum_{j \neq 0} q_{ij}(a_i') |v_n(j) - v_n'(j)|, \quad i \in I - \{0\}.$$

Method of Successive Approximations

Putting the two inequalities together we obtain

$$|v_{n+1}(i) - v'_{n+1}(i)|$$

$$\leq \max\left\{\sum_{j \neq 0} q_{ij}(a_i)|v_n(j) - v_n'(j)|, \sum_{j \neq 0} q_{ij}(a_i')|v_n(j) - v_n'(j)|\right\}$$

for each $n = 0, 1, \ldots$ and $i \in I - \{0\}$. Consequently, letting R_n denote the policy in C_D which takes action a_i or a_i' at state i depending upon which yields the larger value for $\sum q_{ij}(a)|v_n(j) - v_n'(j)|$, we obtain

$$|v_{n+1}(i) - v'_{n+1}(i)| \leq \sum_{j \neq 0} q_{ij}(R_n)|v_n(j) - v_n'(j)|, \quad i \in I - \{0\},$$

for each $n = 0, 1, \ldots$. Repeated iteration yields

$$|v_{n+1}(i) - v'_{n+1}(i)| \leq \sum_{j \neq 0} P_{\tilde{R}_n}\{Y_n = j \mid Y_0 = i\}|v_0(j) - v_0'(j)|$$

$$\leq P_{\tilde{R}_n}\{Y_n \neq 0 \mid Y_0 = i\} \max_j |v(j) - v_0'(j)|,$$

$$i \in I - \{0\},$$

where $\tilde{R}_n$ is the policy in C_M that takes action at time t according to policy R_{n-t} ($0 \leq t \leq n$). From Problem 2, Chapter 3, it follows that

$$\lim_{n \to \infty} P_{\tilde{R}_n}\{Y_n \neq j \mid Y_0 = i\} = 0, \quad i \in I - \{0\}.$$

Therefore

$$\lim_{n \to \infty} |v_n(i) - v_n'(0)| = 0, \quad i \in I - \{0\}.$$

Since $v'_{n+1}(i) = v_0'(i) = \sigma_{R_0}(i)$, $i \in I - \{0\}$, we have that $\lim_{n \to \infty} v_n(i) = \sigma_{R_0}(i)$, $i \in I - \{0\}$, and the theorem is proved.

The remarks regarding the method of successive approximations in Chapter 4 hold here as well. In particular we have:

COROLLARY 1. The function $\sigma_{R_0}(i)$, $i \in I - \{0\}$, uniquely satisfies

$$\sigma_{R_0}(i) = \min_a\left\{w_{ia} + \sum_{j \neq 0} q_{ij}(a)\sigma_{R_0}(j)\right\}, \quad i \in I - \{0\}. \quad (2)$$

Proof: Same as for Corollary 1 to Theorem 1, Chapter 4.

Thus, we start the method of successive approximations with an arbitrary function $v_0(i)$, $i \in I - \{0\}$, and iterate it according to (1). In the limit, we get (2) with $R_0 \in C_D$ as that policy determined by those actions which minimize the right-hand side of (2). In practice, the limit is not attained, but a large number of iterations of (1) should in most cases yield the optimal policy or a good approximation.

Policy Improvement Procedure

We turn now to the policy improvement procedure for obtaining an optimal policy. Let R denote an arbitrary policy in C_D. Then $\{\sigma_R(i)\}$ satisfies uniquely (Theorem 2 of Appendix A) the system

$$\sigma_R(i) = w_{iR} + \sum_{j \neq 0} q_{ij}(R)\sigma_R(j), \qquad i \in I - \{0\}. \tag{3}$$

For each $i \in I - \{0\}$ let E_i denote those actions a for which

$$w_{ia} + \sum_{j \neq 0} q_{ij}(a)\sigma_R(j) < \sigma_R(i).$$

Define $R' \in C_D$ by choosing an action in E_i for at least one i where E_i is not empty. At all other states let $R' = R$. If E_i is nonempty for at least one i we call the transformation of R to R' a *policy improvement iteration*. A sequence of policy improvement iterations that leads to an optimal policy is called the *policy improvement procedure*. That, in fact, every policy improvement procedure leads to an optimal policy is summarized in the following theorem and corollary.

THEOREM 2. If R' is obtained from R by a policy improvement iteration, then $\sigma_{R'}(i) \leq \sigma_R(i)$, $i \in I - \{0\}$, with strict inequality holding for at least one $i \in I - \{0\}$.

Proof: From (3) on substituting the inequalities of the policy

Linear Programming Formulations

improvement iteration, we have

$$\sigma_R(i) \geq w_{iR'} + \sum_{j \neq 0} q_{ij}(R')\sigma_R(j), \qquad i \in I - \{0\},$$

with strict inequality holding for at least one $i \in I - \{0\}$, namely, for i where $R' \neq R$. On iterating the inequality we obtain

$$\sigma_R(i) \geq w_{iR'} + \sum_{j \neq 0} q_{ij}(R') \left[w_{jR'} + \sum_{l \neq 0} q_{jl}(R')\sigma_R(l) \right]$$

$$= w_{iR'} + \sum_{j \neq 0} q_{ij}(R')w_{jR'} + \sum_{j \neq 0} q_{ij}^{(2)}(R')\sigma_R(l)$$

$$\vdots$$

$$= \sum_{t=0}^{T} \sum_{j \neq 0} q_{ij}^{(t)}(R')w_{jR'} + \sum_{j \neq 0} q_{ij}^{(T+1)}(R')\sigma_R(j), \qquad T = 1, 2, \ldots.$$

On letting $T \to \infty$ we obtain

$$\sigma_R(i) \geq \sigma_{R'}(i) + \lim_{T \to \infty} \sum_{j \neq 0} q_{ij}^{(T+1)}(R')\sigma_R(j)$$

$$= \sigma_{R'}(i), \qquad i \in I - \{0\},$$

since $\lim_{T \to \infty} q_{ij}^{(T)}(R') = 0$. Strict equality holds, at least for those i where $R' \neq R$. The theorem is proved.

COROLLARY 1. *The policy improvement procedure converges, within a finite number of policy improvement iterations, to an optimal policy.*

Proof: Each iteration yields a strictly better policy. Only a finite number of policies are in C_D. Thus, at some point, no policy iteration is possible and (2) is satisfied by the final policy; by the corollary to Theorem 1 it must be optimal.

Linear Programming Formulations

We allude now to the linear programming formulations, the primal and dual, of the optimal first-passage problem.

Consider first what we call the primal problem:

To maximize
$$\sum_{j\neq 0} \beta_j v_j$$

subject to
$$v_i \leq w_{ia} + \sum_{j\neq 0} q_{ij}(a) v_j, \qquad a \in K_i, \quad i \in I - \{0\},$$

where the $\{\beta_j\}$ are known positive numbers such that $\sum_{j\neq 0} \beta_j = 1$. The dual problem is:

to minimize
$$\sum_{i\neq 0} \sum_a x_{ia} w_{ia}$$

subject to
$$x_{ia} \geq 0, \qquad a \in K_i, \qquad i \in I - \{0\},$$

and
$$\sum_{i\neq 0} \sum_a x_{ia}(\delta_{ij} - q_{ij}(a)) = \beta_j, \qquad j \in I - \{0\}.$$

By the same methods of Theorem 3, Chapter 4 we can assert that there is a one-to-one correspondence between any solution $\{x_{ia}\}$ to the dual problem, and $R \in C_S$ given by

$$D_{ia}^R = \frac{x_{ia}}{\sum_a x_{ia}}, \qquad a \in K_i, \quad i \in I - \{0\},$$

and

$$x_{ia} = \sum_{l \in I - \{0\}} \beta_l \sum_{t=0}^{\infty} P_R\{Y_t = i, \ Y_n \neq 0, \ 0 \leq n \leq t \mid Y_0 = l\} D_{ia}^R$$
$$= \sum_{l \in I - \{0\}} \beta_l \sum_{t=0}^{\infty} q_{li}^{(t)}(R) D_{ia}^R, \qquad a \in K_i, \quad i \in I - \{0\}.$$

In words, x_{ia} is equal to the expected number of times under the policy R that the process is in state i and action a is taken before the process

Computational Examples

enters state 0 given that $P\{Y_0 = l\} = \beta_l, l \in I - \{0\}$. That x_{ia}, in fact, is finite for $R \in C_S$, follows from the assumption that each $i, i \in I - \{0\}$, is transient for every $R \in C_D$ and Theorem 2 of Appendix A.

Thus, Theorems 3 and 4 of Chapter 4 and their corollaries have their counterpart for the optimal first-passage problem with Eq. (2) of this chapter replacing (3.2) of those discussions. When some β_j's are equal to zero the remark in Chapter 4 holds here as well.

Computational Examples

In order to keep the computations extremely simple we shall consider a two-state problem with one of the states as the target state. Clearly, for such a simple case, the optimal policy can be seen by inspection. However, we shall formally go through the steps of the various procedures.

Suppose $I = \{0, 1\}$; 0 is the target state; $K_1 = 2, q_{11}(1) = \frac{1}{2}, q_{11}(2) = \frac{2}{3}$; $w_{11} = 3, w_{12} = 1$.

We first use the method of successive approximations. Let $v_0(1) = 0$. Then

$$v_1(1) = \min\{3 + \tfrac{1}{2}v_0(1), \quad 1 + \tfrac{2}{3}v_0(1)\}$$
$$= \min\{3, 1\}$$
$$= 1.$$
$$v_2(1) = \min\{3 + \tfrac{1}{2} \cdot 1, \quad 1 + \tfrac{2}{3} \cdot 1\}$$
$$= \tfrac{5}{3}.$$
$$v_3(1) = \min\{3 + \tfrac{1}{2} \cdot \tfrac{5}{3}, \quad 1 + \tfrac{2}{3} \cdot \tfrac{5}{3}\}$$
$$= \tfrac{19}{9}.$$

On the basis of $v_3(1)$, we have that

$$3 + \frac{1}{2}\frac{10}{9} > 1 + \frac{2}{3}\frac{10}{9};$$

hence, the approximation to the optimal policy is to take action $a = 2$

at state 1. In this case the approximation is, in fact, the optimal action.

Using policy improvement, suppose R_1 is the policy which prescribes action $a = 1$ at state 1. Then

$$\sigma_{R_1}(1) = 3 + \tfrac{1}{2}\sigma_{R_1}(1),$$

and, therefore $\sigma_{R_1}(1) = 6$. Since

$$6 > 1 + \tfrac{2}{3}\sigma_{R_1}(1)$$
$$= 5,$$

R_2, which prescribes action $a = 2$, is the policy obtained by the policy iteration; C_D contains only the policies C_1 and C_2; therefore R_2 is optimal. To evaluate $\sigma_{R_2}(1)$, we have that

$$\sigma_{R_2}(1) = 1 + \tfrac{2}{3}\sigma_{R_2}(1),$$

or

$$\sigma_{R_2}(1) = 3.$$

The primal linear programming problem looks like:

Maximize

$$v_1$$

subject to

$$\tfrac{1}{2}v_1 \leqq 3 \quad \text{and} \quad \tfrac{1}{3}v_1 \leqq 1.$$

Clearly the solution is $v_1 = 3$, and since equality is obtained at the second constraint, action $a = 2$ is optimal. The dual problem takes the form:

Minimize

$$3x_{11} + x_{12}$$

subject to

$$x_{11} \geqq 0, \quad x_{12} \geqq 0,$$
$$\tfrac{1}{2}x_{11} + \tfrac{1}{3}x_{12} = 1.$$

The solution is $x_{12} = 3$ which yields $D_{12} = 1$, as the optimal policy.

The Finite Horizon Problem as a First-Passage Problem

The finite horizon problem of Chapter 2 can be viewed as an optimal first-passage problem. Let I' be the state space consisting of all pairs $i' = (i, t)$, $i \in I$, $t = 0, 1, \ldots, T$ and an adjoined state 0 (say); let $K_{i'} = K_i$, $i' \in I' - \{0\}$, $K_0 = 1$, $q_{i'j'}(a) = q_{ij}(a)$ if $i' = (i, t), j' = (j, t+1)$ for $i, j \in I$, $t = 0, 1, \ldots, T-1$, $q_{i'0}(a) = 1$ if $i' \in \{(i, T), i \in I\}$, $q_{00}(a) = 1$ and $q_{i'j'}(a) = 0$ otherwise; $w_{i'a} = w_{ia}$, $i \in I$, $t = 0, \ldots, T$, and $w_{0a} = 0$. Denote by C' the class of all policies. This is merely an enlargement of the original state space to one where the new state designation includes the time of observation as well as the original state; state 0 denotes time $T + 1$ without concern for the original state at time $T + 1$. Within this conception the state 0 is an absorbing state and a first passage from any state in $\{(i, 0), i \in I\}$ to state 0 takes exactly $T + 1$ units of time.

Thus, it should be clear, that to find $R \in C$ to minimize $S_{R,T}(i)$, $i \in I$, is equivalent to finding $R \in C'$ to minimize $\sigma_R((i, 0))$, $i \in I$, where the "target" is the state 0.

This observation coupled with the contents of this chapter point out that $S_{R,T}(i)$ can be minimized by the method of successive approximations, the policy improvement procedure, and by linear programming. This is not to say that any of these methods would be superior to the method of Chapter two. The simple dynamic programming algorithm is the natural and highly efficient way to solve the problem. However, when certain types of additional constraints are imposed the dual linear programming approach may prove useful.

For example, suppose, for a given initial state i, we wish to find $R \in C_S$ to minimize $S_{R,T}(i)$ subject to the constraint that $S_{R,T}(i) \geq s$ (a given constant). Translated to the first-passage problem this would be equivalent to finding $R \in C_S'$ (the stationary Markovian subclass of C') such that $\sigma_R((i, 0))$ is minimized subject to $\sigma_R((i, 0)) \geq s$. Letting $\beta_{(i,0)} = 1$, $\beta_{i'} = 0$, $i' \neq (i, 0)$, this problem can then be formulated as finding $\{x_{i'a}\}$, to minimize

$$\sum_{i' \neq 0} \sum_a x_{i'a} w_{i'a}$$

subject to

$$x_{i'a} \geq 0, \qquad a \in K_{i'}, \quad i' \in I' - \{0\},$$

$$\sum_{i' \neq 0} \sum_a x_{i'a}(\delta_{i'j'} - q_{i'j'}(a)) = \beta_{j'}, \qquad j' \in I' - \{0\}$$

and

$$\sum_{i' \neq 0} \sum_a x_{i'a} w_{i'a} \geq s.$$

If $\{x_{i'a}\}$ is the optimal solution to this linear programming problem, set $D_{i'a} = x_{i'a} / \sum_a x_{i'a}$ if $\sum_a x_{i'a} > 0$ and let $D_{i'a}$ be arbitrary if $\sum_a x_{i'a} = 0$. We point out that the policy so obtained will not in general be a member of C_D' since for at least one state i' a random mechanism will be used for deciding on which action to take.

Bibliographical Remarks

The optimal first-passage problem was formulated by Eaton and Zadeh [30]; they called it a "pursuit problem."

The transformation (1) is not, in general, a contraction for the l^∞ norm. However, because all states except 0 are transient, the proof of the convergence of the method of successive approximations proceeds along the lines of the previous chapter. Other norms are given by Veinott [53] for which (1) is a contraction.

A different linear programming formulation involving the minimization of the ratio of two linear forms (a problem called a fractional linear programming problem which can be readily transformed into a linear programming problem) was first given by Derman [15]. The one given here circumvents the need for the fractional linear programming form.

The remark regarding the formulation of the finite horizon problem as a first-passage problem with application to constrained optimal policies appears in Derman and Klein [22].

Problems

(1) Is it possible to put a bound on $|v_n(i) - \sigma_{R_0}(i)|$, $i \in I - \{0\}$?

(2) For the data given in Problem 1, Chapter 2, find the optimal policy using each method.

6

Expected Average Cost Criterion Computational Procedures

Summary

In Chapters 4 and 5, it was shown that the method of successive approximations, the policy improvement procedure, and linear programming provide general methods for obtaining optimal policies for the discounted cost criterion and for the first-passage problem. In this chapter, which is devoted to the expected average cost criterion, we shall see that a special kind of policy improvement procedure and the methods of linear programming provide general algorithms for obtaining optimal policies.

Policy Improvement Procedure

We first consider the policy improvement procedure. Let $R \in C_D$ be arbitrary. We let

$$\pi_{ij}(R) = \lim_{T \to \infty} \frac{1}{T+1} \sum_{t=0}^{T} q_{ij}^{(t)}(R), \quad i, j \in I,$$

the limit always existing (Theorem 1 of Appendix A). We also have (Theorem 1 of Appendix A):

$$\pi_{il}(R) = \sum_j \pi_{ij}(R) q_{jl}^{(t)}(R) = \sum_j q_{ij}^{(t)}(R) \pi_{jl}(R)$$

$$= \sum_j \pi_{ij}(R) \pi_{jl}(R), \quad i, j \in I, \quad t = 0, 1, \ldots,$$

relations which we shall use throughout.

Consider the equations in $\{\phi_i, v_i, \ldots i \in I\}$:

$$\phi_i + v_i = w_{iR} + \sum_j q_{ij}(R) v_j, \quad i \in I \tag{1}$$

and

$$\sum_j \pi_{ij}(R) v_j = 0, \quad i \in I. \tag{2}$$

Equations (1) and (2) are the essence of the policy improvement procedure. We shall construct a solution to (1) and (2).

Let $E_1, E_2, \ldots, E_k$ be the recurrent classes of I under R. Let $E = \{j_1, \ldots, j_k\}$ be a set of selected states from $E_1, \ldots, E_k$; that is, $j_n \in E_n$, $n = 1, \ldots, k$. Define $w'_{iR} = w_{iR} - \phi_R(i)$, $i \in I$; let $W'_t = w'_{jR}$, if $Y_t = j$, and set $\tau = \min\{t \mid Y_t \in E, t \geq 1\}$. Let $u_R(i) = E_R\{\sum_{t=0}^{\tau-1} W'_t \mid Y_0 = i\}$, $i \in I$; that is, $u_R(i)$ is the expected cost under R and the cost structure $\{w'_{iR}\}$ of going from state i to any of the states in E not counting the cost at the time of arrival.

Policy Improvement Procedure

In constructing a solution to (1) and (2), we first demonstrate that $\{\phi_R(i), u_R(i), i \in I\}$ satisfies the system (1). Then by a suitable modification we can construct a solution to (1) and (2). By its definition we clearly have that

$$u_R(i) = w'_{iR} + \sum_{j \notin E} q_{ij}(R) u_R(j), \qquad i \in I.$$

However, from Theorem 5 of Appendix A, for any $i \in E$ (say $i \in E_n$, where E_n is one of the recurrent classes $E_1, \ldots, E_k$), using the fact that $\pi_{ij} = 0$ if $j \notin E_n$,

$$u_R(i) = \frac{1}{\pi_{ii}(R)} \sum_{j \in I} \pi_{ij}(R) w'_{jR}$$

$$= \frac{1}{\pi_{ii}(R)} \sum_{j \in E_n} \pi_{ij}(R)(w_{jR} - \phi_R(j))$$

$$= \frac{1}{\pi_{ii}(R)} \left(\phi_R(i) - \sum_{j \in E_n} \pi_{ij}(R) \phi_R(j) \right)$$

$$= 0$$

since

$$\sum_{j \in E_n} \pi_{ij}(R) \phi_R(j) = \sum_{j \in E_n} \pi_{ij}(R) \sum_{l \in E_n} \pi_{jl}(R) w_{lR}$$

$$= \sum_{l \in E_n} \pi_{il}(R) w_{lR}$$

$$= \phi_R(i).$$

Therefore,

$$u_R(i) = w'_{iR} + \sum_{j \in I} q_{ij}(R) u_R(j)$$

$$= w_{iR} - \phi_R(i) + \sum_{j \in I} q_{ij}(R) u_R(j), \qquad i \in I,$$

and $\{\phi_R(i), u_R(i), i \in I\}$ is a solution to (1).

Define

$$v_R(i) = u_R(i) - c(i), \qquad i \in I,$$

where

$$c(i) = \sum_{j \in I} \pi_{ij}(R) u_R(j).$$

Since, in fact, $\pi_{ij}(R)$ is independent of i for $i \in E_n$ (Theorem 4 of Appendix A), $c(i)$ is a function of n when i is recurrent. Notice now that

$$\sum_{j \in I} \pi_{ij} v_R(j) = \sum_{j \in I} \pi_{ij}(R)(u_R(j) - c(i))$$

$$= \sum_{j \in I} \pi_{ij}(R) u_R(j) - \sum_{j \in I} \pi_{ij}(R) \sum_{l \in I} \pi_{jl} u_R(l)$$

$$= 0, \quad i \in I$$

since

$$\sum_{j \in I} \pi_{ij}(R) \pi_{jl}(R) = \pi_{il}(R), \quad i, l \in I.$$

Hence, $v_R(i)$ satisfies (2) for all $i \in I$.

Now we also see, using $\sum_j q_{ij}(R) \pi_{jl}(R) = \pi_{il}(R)$, that for $i \in I$,

$$w_{iR} + \sum_{j \in I} q_{ij}(R) v_R(j) = w_{iR} + \sum_{j \in I} q_{ij}(R)(u_R(j) - c(i))$$

$$= w_{iR} + \sum_{j \in I} q_{ij}(R) u_R(j) - c(i)$$

$$= \phi_R(i) + u_R(i) - c(i)$$

$$= \phi_R(i) + v_R(i);$$

that is, $\{\phi_R(i), v_R(i), i \in I\}$ satisfies (1). We have thus constructed a solution $\{\phi_R(i), v_R(i), i \in I\}$ to the combined system (1) and (2). We can further state:

LEMMA 1. The numbers $\{\phi_R(i), v_R(i), i \in I\}$ satisfy (1) and (2); moreover, there is only one solution for which ϕ_i is constant over each recurrent class and equal to $\phi_R(i)$ when i is transient.

Proof: The first statement was just proven. To prove uniqueness, suppose the values $\{\phi_i, v_i, i \in I\}$ satisfy (1) and (2) where ϕ_i is constant on each recurrent class. By premultiplying (1) by $\pi_{li}(R)$ and summing over $i \in I$ and using (2) we have

$$\sum_i \pi_{li}(R) \phi_i = \phi_R(l), \quad l \in I. \tag{3}$$

Policy Improvement Procedure

Suppose l is a recurrent state. Let E_n denote the class of states to which l belongs. Then since π_{li} is independent of l for $i \in E_n$ and $=0$ if $i \notin E_n$, and ϕ_i is constant for $i \in E_n$, it follows from (3) that $\phi_i = \phi_R(i)$ for all $i \in E_n$. Hence, for all recurrent states i, $\phi_i = \phi_R(i)$. Now letting $\Delta_i = v_R(i) - v_i$ and subtracting in (1), we have

$$\Delta_i = \sum_j q_{ij}(R)\, \Delta_j, \quad i \in I.$$

Iterating, we obtain

$$\Delta_i = \sum_j q_{ij}^{(t)}(R)\, \Delta_j, \quad i \in I, \quad t = 1, 2, \ldots \quad (4)$$

Averaging (4) over $t = 1, \ldots, T$ and letting $T \to \infty$, we have by Eq. (2),

$$\Delta_i = \sum_j \pi_{ij}(R)\, \Delta_j$$

$$= 0, \quad i \in I.$$

This proves the lemma.

LEMMA 2. *Given any* α $(0 < \alpha < 1)$ *and* $R \in C_D$,

$$\Psi_R(i, \alpha) = \frac{\phi_R(i)}{(1 - \alpha)} + v_R(i) + \varepsilon_R(i, \alpha), \quad i \in I, \quad (5)$$

where $\varepsilon_R(i, \alpha) \to 0$ *as* $\alpha \to 1$.

Proof: Since $\sum_j q_{ij}(R)\, \phi_R(j) = \phi_R(i)$, it follows on multiplying both sides of (1) by $\alpha^t q_{li}^{(t)}(R)$ and summing over i that

$$\alpha^t \phi_R(l) + \alpha^t \sum_i q_{li}^{(t)}(R) v_R(i)$$

$$= \alpha^t \sum_i q_{li}^{(t)}(R) w_{iR} + \alpha^t \sum_j q_{lj}^{(t+1)}(R) v_R(j), \quad t = 0, 1, \ldots.$$

On summing over t we arrive at

$$(1 - \alpha)^{-1} \phi_R(l) + v_R(l)$$

$$= \Psi_R(l, \alpha) + (1 - \alpha) \sum_{t=1}^{\infty} \alpha^{t-1} \sum_i q_{li}^{(t)}(R) v_R(i), \quad l \in I.$$

Using (2) and the Abelian theorem 1(b) of Appendix B on the last term of the right-hand side, the lemma follows.

Let $R \in C_D$ be arbitrary. For each $i \in I$, define E_i to be the set of actions at state i for which

$$\sum_j q_{ij}(a)\phi_R(j) < \phi_R(i),$$

or, if no actions satisfy the inequality, the set that satisfies

$$\sum_j q_{ij}(a)\phi_R(j) = \phi_R(i)$$

and

$$w_{ia} + \sum_j q_{ij}(a)v_R(j) < w_{iR} + q_{ij}(R)v_R(j)$$
$$= \phi_R(i) + v_R(i).$$

Define $R' \in C_D$ as the policy which takes an action $a \in E_i$ in at least one state i for which E_i is nonempty; otherwise, the action taken is the one dictated by R. Of course, if E_i is empty for all i, then $R = R'$. If $R' \neq R$, then either

$$\sum_j q_{ij}(R')\phi_R(j) \leq \phi_R(i), \qquad i \in I, \qquad (6)$$

with strict inequality holding for at least one i and $q_{ij}(R') = q_{ij}(R)$, $w_{iR} = w_{iR'} j \in I$, for each i where equality holds, or

$$\sum_j q_{ij}(R')\phi_R(i) = \phi_R(i), \qquad i \in I, \qquad (7)$$

and

$$\phi_R(i) + v_R(i) \geq w_{iR'} + \sum_j q_{ij}(R')v_R(j), \qquad i \in I, \qquad (8)$$

with strict inequality in (8) holding for at least one i and $q_{ij}(R') = q_{ij}(R)$, $w_{iR} = w_{iR'} j \in I$, for each i where equality in (8) holds.

LEMMA 3. *If $R' \neq R$, then*

$$\phi_{R'}(i) \leq \phi_R(i), \qquad i \in I, \qquad (9)$$

Policy Improvement Procedure

and

$$\Psi_{R'}(i, \alpha) \leq \Psi_R(i, \alpha), \quad i \in I, \quad \alpha \text{ near } 1, \tag{10}$$

with strict inequality holding in (10) for at least one i.

Proof: From the representation (5) we can write

$$\Psi_R(i, \alpha) = w_{iR} + \alpha \sum_j q_{ij}(R) \left\{ \frac{\phi_R(j)}{1-\alpha} + v_R(j) + \varepsilon_R(j, \alpha) \right\}, \quad i \in I.$$

If (6) holds, for some α_0 near enough to 1 we can write for all $\alpha \geq \alpha_0$,

$$\Psi_R(i, \alpha) \geq w_{iR'} + \alpha \sum_j q_{ij}(R') \left\{ \frac{\phi_R(j)}{1-\alpha} + v_R(j) + \varepsilon_R(j, \alpha) \right\}$$

$$= w_{iR'} + \alpha \sum_j q_{ij}(R') \Psi_R(j, \alpha), \quad i \in I,$$

with strict inequality holding for that i where strict inequality holds in (6). Thus, Theorem 2 of Chapter 4 applies; that is, policy improvement for the discounted cost criterion takes place in going from R to R' for every $\alpha \geq \alpha_0$. If (7) and (8) hold, the same can be said. Thus (10) holds. From the fact that (10) holds and using (5), one sees that (9) also holds. Thus the lemma is proven.

Let us define the transformation from R to R' as a *policy improvement iteration*. Thus, the policy improvement iteration takes a policy $R \in C_D$ to $R' \in C_D$ such that (6) is satisfied or (7) and (8) is satisfied. Thus, the policy improvement iteration is analogous to those discussed in Chapters 4 and 5 though somewhat more complicated. We refer to a sequence of policy improvement iterations as the *policy improvement procedure*. We have:

THEOREM 1. The policy improvement procedure leads to an optimal policy within a finite number of iterations.

Proof: Let $R_1, R_2, \ldots$ be the policies obtained from a sequence of policy improvement iterations with R_1 arbitrary. Since there are only a finite number of policies in C_D and $\{\Psi_{R_n}(i, \alpha), i \in I\}$ is a strictly decreasing sequence as long as a policy improvement iteration can be effected, there is an n for which $R_n = R_{m+1} = R$ (say); that is, a policy iteration on R results in no change of policy. The fact that $\{\Psi_{R_v}(i, \alpha), v = 1, \ldots, n\}$ is strictly decreasing prevents cycling from occurring within the sequence $R_1, \ldots, R_n$. Then we must have

$$\phi_R(i) + v_R(i) = \min_{a \in k_i'} \left\{ w_{ia} + \sum_j q_{ij}(a) v_R(j) \right\}, \qquad i \in I, \quad (11)$$

and

$$\phi_R(i) = \min_a \sum_j q_{ij}(a) \phi_R(j), \qquad i \in I, \quad (12)$$

where K_i' in (11) is the subset of actions at i such that equality is achieved in (12). We now show that whenever R is such that both (11) and (12) hold, then R must be optimal; that is, R is not a local minimum but is, in fact, an absolute minimum. Suppose $\tilde{R}$ is an arbitrary policy in C_D. By virtue of (11) and (12) holding together with the argument employing (5) and its expansion used to prove Lemma 3, we now conclude that

$$\Psi_R(i, \alpha) \leqq w_{i\tilde{R}} + \alpha \sum_j q_{ij}(\tilde{R}) \Psi_R(j, \alpha), \qquad i \in I,$$

for all α sufficiently close to one. By the method of proof used in proving Theorem 2, Chapter 4, we obtain the fact that

$$\Psi_R(i, \alpha) \leqq \Psi_{\tilde{R}}(i, \alpha), \qquad i \in I,$$

for all α sufficiently near 1. From (5) we then conclude that

$$\phi_R(i) \leqq \phi_{\tilde{R}}(i), \qquad i \in I.$$

Since $\tilde{R}$ is arbitrary this proves the theorem.

To spell out the policy improvement procedure, we first start with an arbitrary $R_1 \in C_D$. We then solve for $\{\phi_{R_1}(i), v_{R_1}(i), i \in I\}$. Given $\{\phi_{R_1}(i), i \in I\}$, we obtain $\{v_{R_1}(i), i \in I\}$ algebraically by virtue of Lemma 1. For any $R \in C_D$, $\{\phi_R(i), i \in I\}$ is calculated from $\{\pi_{ij}(R), i, j \in I\}$,

Linear Programming Formulations

where $\{\pi_{ij}(R) = {}_n\pi_j, i, j \in E_n\}$ is the unique solution to

$${}_n\pi_j = \sum_{i \in E_n} {}_n\pi_i \, q_{ij}(R), \qquad j \in E_n,$$

$$\sum_{j \in E_n} {}_n\pi_j = 1$$

(Theorem 4 of Appendix A), and for $i \notin \bigcup_{n=1}^{k} E_n$,

$$\pi_{ij}(R) = \alpha_{in} \, {}_n\pi_j,$$

where $\alpha_{in} = P\{Y_t \in E_n \text{ for some } t \geq 1 \mid Y_0 = i\}$. $\{\alpha_{in}, i \notin \bigcup_{n=1}^{k} E_n\}$ uniquely satisfies the system

$$\alpha_{in} = \sum_{j \in E_n} q_{ij}(R) + \sum_{j \notin \bigcup_{n=1}^{k} E_n} q_{ij}(R) \alpha_{jn}, \qquad i \in \bigcup_{n=1}^{k} E_n.$$

Hence $\{\phi_{R_1}(i), v_{R_1}(i), i \in I\}$ can be obtained algebraically.

Having solved for $\{\phi_{R_1}(i), v_{R_1}(i), i \in I\}$, R_2 is obtained by a policy improvement iteration; that is, at one or more i, where possible, an action a is taken which satisfies either (6) or (7) and (8) with $R_1 = R_2$ at all other states. This process is repeated until (11) and (12) are satisfied, at which point an optimal policy is on hand. Unfortunately, $\phi_R(i)$ and $v_R(i)$ must be obtained anew at each iteration.

Linear Programming Formulations

We now turn to the linear programming approach to obtaining an optimal policy. First consider the linear programming (primal) problem. To determine values of the variables $\{\phi_i, v_i, i \in I\}$

to maximize

$$\sum_j \beta_j \phi_j$$

subject to

$$\sum_j v_j(\delta_{ij} - q_{ij}(a)) + \phi_i \leq w_{ia}, \qquad a \in K_i, \quad i \in I, \qquad (13)$$

$$\sum_j (\delta_{ij} - q_{ij}(a))\phi_j \leq 0, \qquad a \in K_i, \quad i \in I, \qquad (14)$$

where $\beta_j > 0$, $\sum_j \beta_j = 1$ are known constants. The dual problem is to find values of the variables $\{x_{ia}, y_{ia}, a \in K_i, i \in I\}$

to minimize

$$\sum_i \sum_a x_{ia} w_{ia}$$

subject to

$$x_{ia} \geq 0, \qquad y_{ia} \geq 0, \qquad a \in K_i, \quad i \in I,$$

$$\sum_i \sum_a x_{ia}(\delta_{ij} - q_{ij}(a)) = 0, \qquad j \in I, \qquad (15)$$

$$\sum_a x_{ja} + \sum_i \sum_a y_{ia}(\delta_{ij} - y_{ij}(a)) = \beta_j, \qquad j \in I. \qquad (16)$$

We consider the primal problem. In what follows, $R^* \in C_D$ is an optimal policy.

LEMMA. 4. Let $\{\phi_i, v_i, i \in I\}$ be any optimal solution to the primal problem; then $\phi_i = \phi_{R^*}(i)$, $i \in I$.

Proof: From (13) we obtain that

$$\sum_i \pi_{li}(R^*)\phi_i \leq \sum_i \pi_{li}(R^*)w_{iR^*}$$

$$= \phi_{R^*}(l), \qquad l \in I.$$

From (14) we have that for $t = 1, 2$,

$$\sum_j q_{ij}^{(t)}(R^*)\phi_j \geq \phi_l, \qquad l \in I.$$

Hence,

$$\sum_j \pi_{lj}(R^*)\phi_j \geq \phi_l, \qquad l \in I.$$

Thus $\phi_{R*}(l) \geq \phi_l, l \in I$. However, from Eqs. (11) and (12), we see that $\{\phi_{R*}(i), v_{R*}(i) + c\phi_{R*}(i), i \in I\}$ for c large enough is a feasible solution to the primal linear programming problem from which it follows that $\phi_j = \phi_{R*}(j), j \in I$.

LEMMA 5. Let $R \in C_D$ be any optimal policy and $\{\phi_i, v_i, i \in I\}$ an optimal solution to the primal problem; then

$$\sum_j v_j(\delta_{ij} - q_{ij}(R)) + \phi_i = w_{iR}$$

for every i that is recurrent with respect to R, and

$$\sum_j (\delta_{ij} - q_{ij}(R))\phi_j = 0$$

for every $i \in I$.

Proof: The first assertion follows from (13) on premultiplying the appropriate inequality in (13) by $\pi_{li}(R)$ and summing over i. If equality fails to hold, we have $\sum_i \pi_{li}(R)\phi_i < \phi_R(l)$ for some $l \in I$; hence, $\sum_i \pi_{li}(R)\phi_i < \sum_i \pi_{li}(R)\phi_R(i)$, contradicting Lemma 4 and the assumption that R is optimal. The second assertion must hold since, by Lemma 4, $\phi_i = \phi_{R*}(i) = \phi_R(i), i \in I$, and therefore,

$$\sum_j q_{ij}(R)\phi_j = \sum_j q_{ij}(R)\phi_R(j)$$
$$= \phi_R(j)$$
$$= \phi_i, \qquad i \in I.$$

Lemma 5 asserts that in any optimal solution to the primal problem one can always select actions $a = a_i$ for each $i \in I$ such that $\sum_j (\delta_{ij} - q_{ij}(a_i))\phi_j = 0, i \in I$, and $\sum_j v_j(\delta_{ij} - q_{ij}(a_i)) + \phi_i = w_{ia_i}$ for all i in a nonempty subset A of I. Let $R \in C_D$ denote the policy that takes action a_i for $i \in I$.

THEOREM 2. If the states $i \in I - A$ are transient with respect to R then the policy R is optimal.

Proof: By hypothesis, the states i for which $\sum_{j} v_j(\delta_{ij} - q_{ij}(R)) + \phi_i < w_{iR}$ are transient with respect to R. From (14) we have

$$\sum_{i} \pi_{li}(R)\phi_i = \phi_R(l), \quad l \in I,$$

and from (14) with equality holding,

$$\phi_l = \sum_{j} \pi_{lj}(R)\phi_j, \quad l \in I.$$

Hence $\phi_R(l) = \phi_l = \phi_{R^*}(l)$, $l \in I$, and R is optimal.

COROLLARY 1. If for some R there exists $\{v_i, i \in I\}$ satisfying

$$\sum_{j} v_j(\delta_{ij} - q_{ij}(R)) + \phi_{R^*}(i) = w_{iR}, \quad i \in I,$$

$$\sum_{j} (\delta_{ij} - q_{ij}(R))\phi_{R^*}(j) = 0,$$

then R is optimal.

The linear programming method as suggested by Theorem 2 is to solve the primal linear programming problem and choose $R \in C_D$, if possible, by taking those actions for which equality in (13) and (14) are simultaneously obtained. There may be some states where equality in (13) is not attained for any action. If we are fortunate in our selection of actions, then the states where equality is not attained will be transient, in which case R is optimal. However, we may not be so fortunate in our selection of actions as to have the states transient where equality is not attained. We now show that the solution of a second linear programming leads to an optimal policy. The idea behind the method is that of finding new values of the variables $\{v_i\}$ which will force equality in (13) and (14) for at least one decision at every state.

Let $\{\phi_i, v_i, i \in I\}$ be an optimal solution to the primal problem (we refer to this problem as Problem 1). Let A denote the states for which equality in (13) and (14) is achieved for at least one action. Then by Lemma 5, $I - A$ must consist entirely of transient states under every optimal policy. Let A' be the largest subset of A such that for some action a_i satisfying the equality in (13) and (14), we have $q_{ij}(a_i) = 0$ for all $j \in I - A$. By Lemma 5 the states in $A - A'$ must also be transient under an optimal R, because if $i \in A - A'$ is recurrent under an optimal policy, then there exists an action a_i (viz. the action in the optimal policy) for which the equality in (13) and (14) must hold and, since the states of $I - A$ are transient, $q_{ij}(a_i) = 0, j \in I - A$, which is contrary to the definition of A'.

Let $T = I - A'$ and $\tilde{K}_i$ denote the actions at state i, $i \in T$, for which equality in (14) holds. Consider Problem 2: To find $\{u_i, i \in T\}$ to maximize

$$\sum_{i \in T} u_i$$

subject to

$$\sum_{j \in T} u_j(\delta_{ij} - q_{ij}(a)) \leq w_{ia} - \sum_{j \in A'} v_j(\delta_{ij} - q_{ij}(a)) - \phi_i$$

$$= b_i(a), \qquad a \in \tilde{K}_i, \quad i \in T.$$

We shall show that by solving for $\{u_i, i \in T\}$ and replacing v_i by u_i for $i \in T$, we shall be in a position to make use of Corollary 1 of Theorem 2.

LEMMA 6. A finite optimal solution to Problem 2 exists.

Proof: From Lemma 5 it follows that there is at least one set of actions $a_i \in \tilde{K}_i, i \in T$, for which the states $i \in T$ are transient. Let $\{a_i, i \in T\}$ be such a set of actions. Then $\sum_{j \in T} u_j(\delta_{ij} - q_{ij}(a_i)) = b_i(a_i)$, $i \in T$, has a unique solution (Theorem 2 of Appendix A), say $\{u_j^*, j \in T\}$. Let $\{u_j, j \in T\}$ be any solution to the inequalities of Problem 2. In particular we must have

$$\sum_{j \in T} u_j(\delta_{ij} - q_{ij}(a_i)) \leq b_i(a_i), \qquad i \in T,$$

On subtracting, we get

$$\sum_{j \in T} (u_j^* - u_j)(\delta_{ij} - q_{ij}(a_i)) \geq 0, \qquad i \in T,$$

from which (since the inverse of $I - Q$ consists of all positive terms)

$$u_j^* \geq u_j, \quad j \in T,$$

and, hence, $\sum_{j \in T} u_j^* \geq \sum_{j \in T} u_j$.

LEMMA 7. Let $\{u_j, j \in T\}$ be any optimal solution to Problem 2 and $\{a_i, i \in T\}$ be the actions from which

$$\Delta_i(a) = b_i(a) - \sum_{j \in T} u_j(\delta_{ij} - q_{ij}(a))$$

is minimized; then $\Delta_i(a_i) = 0$, $i \in T$.

Proof: If $\Delta_i(a_i) > 0$, then $u_i + \varepsilon$ with $\varepsilon > 0$ and small enough would imply $\{u_j' = u_j, j \neq i, u_i' = u_i + \varepsilon\}$ is also a feasible solution. However, $\{u_j\}$ would not then be optimal.

We now have

THEOREM 3. Let R be the policy obtained by taking the actions dictated by the solutions to Problems 1 and 2; then R is optimal.

Proof: The theorem follows from Lemma 4, Lemma 7, and Corollary 1 of Theorem 2.

Policy Improvement, Linear Programming under Irreducibility Assumption

We now discuss the problem of finding the optimal policy under the assumption (A): I is irreducible for every $R \in C_D$; that is, for every $R \in C_D$, every i is recurrent and every pair of states i and j communicate.

LEMMA 8. If (A) holds, then I is irreducible for every $R \in C_S$.

Proof: Let $R \in C_S$ be arbitrary and $p_{ij} = \sum_a D_{ia}^R q_{ij}(a)$, $i, j \in I$. For

Policy Improvement, Linear Programming

some $a_i, p_{ij} \geq D_{ia_i}^R q_{ij}(a_i)$, where $D_{ia_i}^R > 0, i \in I$. Let $\delta = \min_i \{D_{ia_i}\}$. Hence, $p_{ij} \geq \delta q_{ij}(a_i); i, j \in I$. Let $R_0 \in C_D$ be defined as the policy that takes action $a = a_i$ in state $i, i \in I$. Since I is irreducible under R_0, for each $i, j \in I$ there exists an n such that $q_{ij}^{(n)}(R_0) > 0$. Consequently, $p_{ij}^{(n)} > 0$ for the same value of n. Thus under R every state communicates with every other state from which it follows that I is irreducible under R.

If (A) holds, $\phi_R(i) = \phi_R$ independent of i for every $R \in C_D$ (Theorem 4 of Appendix A). Consequently, (7) always holds so that the policy improvement procedure is involved only with inequalities (8). Also since $\pi_{ij}(R) = \pi_j(R)$ independent of i, Eqs.(2) reduce to a single equation. We can then alternatively replace (2) by the equation; for example, $v_R(j) = 0$ for some given $j \in I$. Thus, the policy improvement procedure takes the form of starting with an initial policy $R_1 \in C_D$ and solving for $\{\phi_{R_1}, v_{R_1}(i), i \in I\}$. Then R_2 is taken to be any policy in C_D that takes action a for at least one state i which reduces $w_{ia} + \sum_j q_{ij}(a)v_{R_1}(j)$. Where reduction cannot or is not effected, the action a under R_1 is taken. We then repeat this process to obtain $R_2, R_3, \ldots$ until for some n, $R_n = R_{n+1}$, at which point R_n is optimal.

The linear programming formulation (the primal problem) is to find variables $\{\phi, v_i, i \in I\}$

to maximize

$$\phi$$

subject to

$$\sum_j v_j(\delta_{ij} - q_{ij}(a)) + \phi \leq w_{ia}, \quad a \in K_i, \quad i \in I. \qquad (17)$$

An optimal solution $\{\phi, v_i, i \in I\}$ will have $\phi = \phi_{R*}$ where R^* is an optimal policy and an optimal policy is obtained from the linear programming solution by taking an action $a = a_i$ in state i where equality holds in (17). Since, by assumption (A) there will be no transient states, an action $a = a_i$ in state i where equality holds in (17), will exist for every $i \in I$.

The dual problem becomes that of finding variables $\{x_{ia}, a \in K_i, i \in I\}$ which

minimize

$$\sum_i \sum_a x_{ia} w_{ia}$$

subject to

$$x_{ia} \geq 0, \qquad a \in K_i, \quad i \in I,$$

$$\sum_i \sum_a x_{ia}(\delta_{ij} - q_{ij}(a)) = 0, \qquad j \in I, \tag{18}$$

$$\sum_i \sum_a x_{ia} = 1.$$

Since under (A), $\{\pi_j(R), j \in I\}$ satisfy uniquely the steady-state equations

$$\sum_i \pi_i(R)(\delta_{ij} - q_{ij}(R)) = 0, \qquad j \in I,$$

$$\sum_j \pi_j(R) = 1,$$

where $\pi_j(R) > 0, j \in I$, by arguments similar to those employed in Theorem 3, Chapter 4, we can assert that there is a one-to-one correspondence between the solutions $\{x_{ia}\}$ to (18) and policies $R \in C_S$ given by

$$D_{ia}^R = \frac{x_{ia}}{\sum_a x_{ia}}, \qquad a \in K_i, \quad i \in I,$$

and

$$x_{ia} = \pi_i(R) D_{ia}^R, \qquad a \in K_i, \quad i \in I.$$

Thus, *every* solution $\{x_{ia}\}$ to the dual problem is capable of a policy interpretation; in particular, $\phi_R = \sum_i \sum_a x_{ia} w_{ia}$. If, in fact, an optimal solution is obtained by the simplex method, then the corresponding policy R will be a member of C_D since at least one of the equations in (18) is redundant and $\sum_a x_{ia} > 0, i \in I$. Hence, for exactly one $a = a_i$, $x_{ia} > 0$ for each state i.

Computational Example

In the next chapter we shall be concerned with obtaining optimal policies under certain types of additional constraints. Here optimal policies will be outside the class C_D. It will turn out to be most useful to approach this kind of problem from the point of view given by the correspondence between x_{ia} and D_{ia} as suggested by the dual problem.

Computational Example

Suppose we have $I = \{0, 1\}$, $K_i = 2$, $i = 0, 1$, where

$$\begin{Bmatrix} w_{01} & w_{02} \\ w_{11} & w_{12} \end{Bmatrix} = \begin{Bmatrix} 1 & 3 \\ 4 & 0 \end{Bmatrix}$$

and

$$\begin{Bmatrix} (q_{00}(1), q_{00}(2)) & (q_{01}(1), q_{01}(2)) \\ (q_{10}(1), q_{10}(2)) & (q_{11}(1), q_{11}(2)) \end{Bmatrix} = \begin{Bmatrix} (1, \tfrac{1}{2}) & (0, \tfrac{1}{2}) \\ (1, 0) & (0, 1) \end{Bmatrix}.$$

First, we use the policy improvement procedure to find the optimal policy. Let R_1 be the policy that takes action $a = 1$ at state 0 and action $a = 1$ at state 1. The transition matrix under R_1 is

$$\begin{Bmatrix} q_{01}(R_1) & q_{01}(R_1) \\ q_{11}(R_1) & q_{11}(R_1) \end{Bmatrix} = \begin{Bmatrix} 1 & 0 \\ 1 & 0 \end{Bmatrix}.$$

Clearly $\pi_{00}(R_1) = \pi_{10}(R_1) = 1$ and $\pi_{01}(R_1) = \pi_{11}(R_1) = 0$, from which $\phi_{R_1}(0) = \phi_{R_1}(1) = 1$. Equations (1) become

$$1 + v_0 = 1 + v_0,$$
$$1 + v_1 = 4 + v_0,$$

from which $v_0 = v_1 - 3$. From (2) we have $v_0 = 0$; hence, $v_1 = 3$. Now

$$3 + \tfrac{1}{2}v_0 + \tfrac{1}{2}v_1 > 1$$

and

$$v_1 = 3 < 4;$$

therefore R_2, action $a = 1$ at state 0 and action $a = 2$ at state 1, is an improvement over R_1. The matrix of transition probabilities under R_2

is

$$\begin{pmatrix} q_{00}(R_2) & q_{01}(R_2) \\ q_{10}(R_2) & q_{11}(R_2) \end{pmatrix} = \begin{pmatrix} 1 & 0 \\ 0 & 1 \end{pmatrix}.$$

This time $\pi_{00}(R_2) = 1$, $\pi_{01}(R_2) = 0$, $\pi_{10}(R_2) = 0$, $\pi_{11}(R_2) = 1$, where $\phi_{R_2}(0) = 1$, $\phi_{R_2}(1) = 0$. Now

$$q_{00}(a_2)\phi_{R_2}(0) + q_{01}(a_2(\phi_{R_2}(1) = \tfrac{1}{2}\phi_{R_2}(0) = \tfrac{1}{2}\phi_{R_2}(1)$$
$$= \tfrac{1}{2}$$
$$< \phi_{R_2}(0) = 1;$$

hence R_3, action $a = 2$ at state 0 and action $a = 2$ at state 1, is an improvement over R_2. The matrix of transition probabilities under R_3 is

$$\begin{pmatrix} q_{00}(R_3) & q_{01}(R_3) \\ q_{10}(R_3) & q_{11}(R_3) \end{pmatrix} = \begin{pmatrix} \tfrac{1}{2} & \tfrac{1}{2} \\ 0 & 1 \end{pmatrix}.$$

Here we must have $\pi_{00}(R_3) = \pi_{10}(R_3) = 0$ since state 0 is transient under R_3, while $\pi_{01}(R_3) = \pi_{11}(R_3) = 1$. Thus, $\phi_{R_3}(0) = 0$, $\phi_{R_3}(1) = 0$. Equations (1) become

$$v_0 = 3 + \tfrac{1}{2}v_0 + \tfrac{1}{2}v_1$$
$$v_1 = v_1;$$

hence, $v_0 = 6 + v_1$. From Eq. (2), $v_1 = 0$; thus $v_0 = 6$. Now

$$w_{01} + q_{00}(1)v_0 + q_{01}(1)v_1 = 1 + 6$$
$$> \phi_{R_3}(0) + v_0$$
$$= 6$$

and

$$w_{11} + q_{10}(1)v_0 + q_{11}(1)v_1 = 4$$
$$> \phi_{R_3}(1) + v_1$$
$$= 0.$$

Therefore, R_3 is optimal.

We can write the primal linear programming problem: to find ϕ_0, ϕ_1, v_1, and v_2

to maximize
$$\tfrac{1}{2}\phi_0 + \tfrac{1}{2}\phi_1$$
subject to
$$\phi_0 \leqq 1,$$
$$\tfrac{1}{2}v_0 - \tfrac{1}{2}v_1 + \phi_0 \leqq 3,$$
$$\phi_1 \leqq 4,$$
$$\phi_1 \leqq 0,$$
$$\tfrac{1}{2}\phi_0 - \tfrac{1}{2}\phi_1 \leqq 0,$$
$$\phi_0 \leqq 0.$$

By inspection, it is seen that an optimal solution is
$$\phi_0 = \phi_1 = 0, \qquad v_0 = v_1 = 1.$$
Equality in (13) is achieved at state 1 with action $a = 2$. Equality in (14) is achieved by actions $a = 1$ and 2 at state 0 and by action $a = 2$ in state 1. Problem 2 becomes: choose u_0

to maximize
$$u_0$$
subject to
$$\tfrac{1}{2}u_0 \leqq 3 + \tfrac{1}{2};$$
that is, $u_0 = 7$. Setting $v_0' = 7$, $v_1' = 1$, we get from Corollary 1 of Theorem 2 that R, action $a = 2$ at both states 0 and 1, is optimal.

Bibliographical Remarks

The policy improvement procedure is due to Howard [34].

The proof of the existence and uniqueness of solutions to Eqs. (1) and (2) given here stems from the approach taken by Derman and Veinott [26]. Blackwell [6] originally gave a different development of representation (5) involving solutions to (1) and (2). Equations like (11) and (12) appear in Bellman [2]. Miller and Veinott [47] and Veinott [53] generalize the representation (5); that is they are able to express $\Psi_R(i, \alpha)$ in a Laurent expansion in $\rho = (1 - \alpha)/\alpha$ (the interest rate when α is the discount factor). Just as (5) is crucial to the policy improvement procedure, the Laurent expansion in ρ is used to obtain algorithms for finding policies optimal to more sensitive criteria. In particular, an algorithm can be given to obtain a policy optimal in the sense of Corollary 1 to Theorem 1 of Chapter 3. Also in [53] a more efficient method for computing $\{\phi_R(i), v_R(v), i \in I\}$ is given which does not require computation of the quantities $\{\pi_{ij}(R)\}$.

The linear programming method in the dual form was given first by Manne [46] for the case where all states belong to one irreducible class for every policy $R \in C_D$. The multiple class case is due to Denardo and Fox [13], although the first attempted linear programming approach utilizing the first of the two problems goes back to Balinski [1]. No satisfactory treatment of the dual problem for the multiple class case has been published.

Problems

(1) Suppose P is any stochastic matrix and

$$\Pi = \lim_{N \to \infty} \frac{1}{N} \sum_{n=1}^{N} P^n.$$

(a) Show that the rank of $\{{}^I_\Pi{}^{-P}\}$ is equal to the number of rows of P.
(b) Prove that $I - P + \Pi$ has an inverse.
(c) (Veinott [53]). Show that if B is a square matrix for which $\lim_{N \to \infty} \sum_{n=0}^{N} B^n/(N + 1) = 0$, then $I - B$ is nonsingular and $(I - B)^{-1} = \lim_{N \to \infty} \sum_{n=0}^{N} \sum_{k=0}^{n} B^k/(N + 1)$.
Hint: $(I - B) \sum_{n=0}^{N} B^n = I - B^{N+1}$ for every N.

(d) As an alternative approach use (c) to prove (b) giving a representation for $(I - P + \Pi)^{-1}$.

(2) For the data provided in Chapter 2, p. 17, solve for the optimal average cost policy by policy improvement and by linear programming.

(3) Let the costs be as in Problem 2. However, suppose

$$\begin{Bmatrix} (q_{00}(1), q_{00}(2)) & (q_{00}(1), q_{01}(2)) & (q_{02}(1), q_{02}(1)) \\ (q_{10}(1), q_{10}(2)) & (q_{11}(1), q_{11}(2)) & (q_{12}(1), q_{12}(2)) \\ (q_{20}(1), q_{20}(2)) & (q_{21}(1), q_{21}(2)) & (q_{22}(1), q_{22}(2)) \end{Bmatrix}$$

$$= \begin{Bmatrix} (\tfrac{1}{2}, \tfrac{1}{3}) & (\tfrac{1}{2}, \tfrac{1}{3}) & (0, \tfrac{1}{3}) \\ (\tfrac{1}{4}, \tfrac{1}{2}) & (\tfrac{3}{4}, 0) & (0, \tfrac{1}{2}) \\ (0, \tfrac{1}{2}) & (0, \tfrac{1}{4}) & (1, \tfrac{1}{4}) \end{Bmatrix};$$

find the optimal average cost policy using the policy improvement procedure and linear programming.

(4) Show that Lemma 1 does not necessarily hold if $\phi_i \neq \phi_R(i)$ when i is transient.

(5) Show that the optimal set of actions to be taken in the states of T (defined for Problem 2 in the linear programming formulation) are any set which make all the states of T transient.

(6) Given any subset I' of I, for which I' is inaccessible from the states of $I - I'$, construct an algorithm that will find a policy $R \in C_D$, if one exists, which has the property that all the states in I' are transient.

7

State-Action Frequencies and Problems with Constraints

Introduction and Summary

   Most problems encountered involve cost criteria expressible in terms of the frequencies of the occurrence of the various combinations of states and actions as the decision process evolves over time. In seeking an optimal policy, we have seen in the cases studied that not all policies have to be considered. In the problems of Chapters 4, 5, and 6 it is sufficient to limit consideration to the policies of C_D although for computational purposes, we did allow policies from C_S. The aim of this chapter is to show that the expected long range state-action frequencies generated by a decision process under an arbitrary policy can be reproduced by a policy belonging to C_M, or, under certain conditions, to C_S or C_D. Once having shown this, we will possess a most useful tool for enabling us to assert that optimal policies for other problems exist somewhere in the class C_M, C_S, or C_D. An analogous result will be obtained for the long range state-action frequencies without taking expectations.

This approach becomes particularly useful when optimization problems involving nonlinear functions or side constraints are considered. In such problems, it is not obvious (nor necessarily true) that optimal policies exist at all, or if they do that one can be found in the class C_D or C_S. For example, suppose 0 is the initial state of a dynamic system and j is a state to be avoided if practicable; for example, j may represent a state at which the system is inoperable. Only one action is available in state j and $q_{j0}(1) = 1$. In other words, when the system fails it takes a unit of time to make it operable, after which the system is always returned to its initial state 0. However, there are other states from which the system may be returned to 0. Each return to 0 is costly, although the costs may not be assessable. All that can be said is that returns to 0 from failure are more costly than returns from nonfailure states. Under these circumstances a not unreasonable problem can be formulated: to find the policy $R \in C$ that maximizes the expected time until a recurrence of state 0 takes place, subject to the constraint, that $P\{Y_t = j, t < \tau \mid Y_0 = 0\} \leq \alpha$, where $\tau = \min\{t \mid Y_t = 0, t \geq 1\}$, and α is a given number between 0 and 1. At this point, it is not obvious that an optimal policy must exist for this problem which is a member of the class C_D or C_S. It will be shown later that the expected recurrence time of state 0, $E\tau$, and the constraint can be expressed in terms of expected state-action frequencies and that an optimal policy exists in C_S.

Expected State-Action Frequencies

With some apologies for the notation, let

$$x_{Tja}^R(i) = \frac{1}{T+1} \sum_{t=0}^{T} P_R\{Y_t = j, \quad A_t = a \mid Y_0 = i\};$$

in words, under policy R and given $Y_0 = i$, $x_{Tja}^R(i)$ is the expected frequency, up to time T, of entrances into state j where action a is taken. Let $X_T^R(i)$ denote the matrix (we assume, with no loss of generality, $K_i = K$ for all i) of $X_{Tja}^R(i)$ over all $a \in K_i, j \in I$.

For a given policy R, let $H_R(i)$ denote the set of all limit points of the sequence $\{X_T^R(i), T = 0, 1, \ldots\}$. Under some policies, at least those in C_S, $H_R(i)$ will consist of one point. However, for the larger classes this may not be true. Let

$$H(i) = \bigcup_{R \in C} H_R(i),$$

$$H^M(i) = \bigcup_{R \in C_M} H_R(i),$$

$$H^S(i) = \bigcup_{R \in C_S} H_R(i),$$

$$H^D(i) = \bigcup_{R \in C_D} H_R(i).$$

Thus, for example, $H(i)$ is the totality of limit points of $\{X_T^R(i), T = 0, 1, \ldots\}$ obtainable as R assumes all possible policies.

Further, we let $\bar{H}^S(i)$ and $\bar{H}^D(i)$ denote the respective closed convex hulls of $H^S(i)$ and $H^D(i)$.

Some Examples

The following are examples of problems, where the cost criterion and constraints are expressible as functions of points in $H(i)$ or subsets thereof.

Example 1. To minimize $\phi_R(i)$, the expected average cost. We need only note that

$$\phi_R(i) = \lim_{T \to \infty} \inf \text{ (or lim sup)} \sum_j \sum_a x_{Tja}^R(i) w_{ja}.$$

Thus, for some point $X^R(i) = \{x_{ja}^R(i)\} \in H_R(i)$,

$$\phi_R(i) = \sum_j \sum_a x_{ja}^R(i) w_{ja}$$

and for some point $\{X_{ja}(i)\} \in H(i)$ (or, as we have shown in $H^D(i)$), ϕ_R is minimized.

Example 2. Suppose $Y_0 = i$. Let $\tau = \min\{t \mid Y_t = i, t \geq 1\}$. Assume $E_R \tau < \infty$ for every $R \in C$. Let

$$\sigma_R(i) = E_R \sum_{t=1}^{\tau} W_t = \sum_j \sum_a \eta_{ja} w_{ja},$$

where η_{ja} is the expected number of t such that $Y_t = j$ and $A_t = a$ for $1 \leq t \leq \tau$. The problem is to minimize $\sigma_R(i)$. Since the cost criterion $\sigma_R(i)$ is involved with the decision only from $t = 0$ until $t = \tau$, the first time of reentry to state i, we need only consider the class of policies that " begin again" every time i is entered; we call these "renewal policies." Under such policies $\{Y_t\}$ is a "recurrent event" process (see Appendix B), entry into state i being the recurrent event. For such processes, from Theorem 6 of Appendix B,

$$\lim_{T \to \infty} x_{Tja}(i) = \frac{\eta_{ja}}{E\tau}, \quad \text{if} \quad E\tau < \infty.$$

However, $\sum_a \eta_{ia} = 1$ since there is precisely one entry into i for $1 \leq t \leq \tau$. Hence,

$$E\tau = \left(\lim_{T \to \infty} \sum_a x_{Tia}(i) \right)^{-1}.$$

(Note, the equation extends to the case where $E\tau = \infty$; that is, if $E\tau = \infty$, then $\lim_{T \to \infty} \sum_a x_{Tia}(i) = 0$). Therefore,

$$\sigma_R(i) = \frac{\sum_j \sum_a \lim_{T \to \infty} x_{Tja}^R(i) w_{ja}}{\lim_{T \to \infty} \sum_a x_{Tia}^R(i)}$$

$$= \lim_{T \to \infty} \frac{\sum_j \sum_a x_{Tja}^R(i) w_{ja}}{\sum_a x_{Tia}^R(i)}$$

so that $\sigma_R(i)$ is expressible as a function of the points in $H(i)$. In particular, the problem is to find the point $\{x_{ja}^R(i)\}$ in $\bigcup_{R \in C'} H_R(i)$ which minimizes $\sigma_R(i)$, where C' is the class of renewal policies.

The Main Theorems and Applications

Example 3. Suppose $Y_0 = i$ and τ is as in Example 2. Let $j_r, \ldots, j_r \neq i$ be such that $q_{j_v i}(a) \equiv 1, v = 1, \ldots, r$. Let

$$\alpha_{j_v} = P\{Y_t \neq j_v, \ 1 \leq t \leq \tau \mid Y_0 = i\}, \qquad v = 1, \ldots, r.$$

Consider the problem of maximizing $E\tau$ subject to constraints

$$\alpha_{j_v} \leq \alpha_v, \qquad v = 1, \ldots, r.$$

In Example 2 it was shown that $E\tau$ is expressible as a function of points in $H(i)$. Since $Y_t = j_v$, $1 \leq t \leq \tau$, for at most one t, it follows that

$$\alpha_{j_v} = 1 - \sum_a \eta_{j_v a}, \qquad v = 1, \ldots, r.$$

However, since η_{ja} are expressible as functions of points in $H(i)$, the problem can be expressed as finding that point in $H(i)$ which maximizes $E\tau$ subject to the given constraints.

The Main Theorems and Applications

We now proceed to the main theorems.

THEOREM 1. *Let* $Y_0 = i, R_1, \ldots, R_n \in C, \{\beta_v\} \ni \beta_v \geq 0, \sum_{v=1}^n \beta_v = 1.$ *Then there exists an* $R \in C_M$ *satisfying*

$$X_T^R(i) = \sum_{v=1}^n \beta_v X_T^{R_v}(i), \qquad \text{for all} \quad T.$$

Proof: Consider a policy generated by selecting R_v ($v = 1, \ldots, n$) at random according to selection probabilities $\beta_v(v = 1, 2, \ldots, n)$; that is, introduce an initial randomization over the policies $R_1, \ldots, R_n$. Denote this policy by $\tilde{R}$. Strictly speaking, $\tilde{R}$ is outside of the class C of all rules which we are considering since $\tilde{R}$ not only depends on the history of the process but also the outcome of the initial randomization. However, $\tilde{R}$ is a policy from the use of which $\{Y_t, A_t, t = 0,$

1, ...} is a stochastic process. Define

$$D_{ja}^R(t) = P_R\{A_t = a \mid Y_t = j\}$$
$$= P_{\bar{R}}\{A_K = a \mid Y_t = j, \quad Y_0 = i\}$$
$$= \frac{P_{\bar{R}}\{Y_t = j, \quad A_t = a \mid Y_0 = i\}}{P_{\bar{R}}\{Y_t = j \mid Y_0 = i\}}$$
$$= \frac{\sum_{v=1}^{n} \beta_v P_R\{Y_t = j, \quad A_t = a \mid Y_0 = i\}}{\sum_{v=1}^{n} \beta_v P_{R_v}\{Y_t = j \mid Y_0 = i\}}.$$

We shall show for every $t = 0, 1, \ldots$ that

$$P_R\{Y_t = j, \quad A_t = a \mid Y_0 = i\}$$
$$= P_{\bar{R}}\{Y_t = j, \quad A_t = a \mid Y_0 = i\}$$
$$= \sum_{v=1}^{n} \beta_v P_{R_v}\{Y_t = j, \quad A_t = a \mid Y_0 = i\}, \qquad a \in K_j, \quad j \in I. \quad (1)$$

For $t = 0$ and $j \neq i$, Eq. (1) holds trivially since both sides vanish. For $t = 0$ and $j = i$,

$$P_{\bar{R}}\{Y_0 = i, \quad A_0 = a \mid Y_0 = i\} = \sum_{v=1}^{n} \beta_v P_{R_v}\{Y_0 = i, \quad A_0 = a \mid Y_0 = i\}$$
$$= \frac{\sum_{v=1}^{n} \beta_v P_{R_v}\{Y_0 = i, \quad A_0 = a \mid Y_0 = i\}}{\sum_{v=1}^{n} \beta_v P_{R_v}\{Y_0 = i \mid Y_0 = i\}}$$
$$= P_{\bar{R}}\{A_0 = a \mid Y_0 = i\}$$
$$= P_R\{Y_0 = i, \quad A_0 = i \mid Y_0 = i\}.$$

Hence (1) is true for $t = 0$. Assume (1) is true for $t = 0, \ldots, T - 1$. Now

$$P_R\{Y_T = j, \quad A_T = a \mid Y_0 = i\}$$
$$= P_R\{Y_T = j \mid Y_0 = i\} P_R\{A_T = a \mid Y_T = j, \quad Y_0 = i\}$$
$$= P_R\{Y_T = j \mid Y_0 = i\} D_{ja}^R(t).$$

The Main Theorems and Applications

However, by the induction hypothesis

$$P_R\{Y_T = j \mid Y_0 = i\} = \sum_l \sum_a P_R\{Y_{T-1} = l, \ A_{T-1} = a \mid Y_0 = i\} q_{lj}(a)$$

$$= \sum_l \sum_a P_{\bar{R}}\{Y_{T-1} = l, \ A_{T-1} = a \mid Y_0 = i\} q_{lj}(a)$$

$$= P_{\bar{R}}\{Y_T = j \mid Y_0 = i\}.$$

Hence,

$$P_R\{Y_T = j, \ A_T = a \mid Y_0 = i\} = P_R\{Y_T = j \mid Y_0 = i\} D_{ja}^R(t)$$

$$= P_{\bar{R}}\{Y_T = j \mid Y_0 = i\} D_{ja}^R(t)$$

$$= P_{\bar{R}}\{Y_T = j, \ A_T = a \mid Y_0 = i\},$$

and the induction argument is complete.

COROLLARY 1. Let $H_{\bar{R}}(i)$ be the set of limit points of $\{X_T^{\bar{R}}(i), T = 0, 1, \ldots\}$; then there exists an $R \in C_M$ such that $H_R(i) = H_{\bar{R}}(i)$.

Proof: Since $X_T^{\bar{R}}(i) = X_T^R(i)$ for every T if R is the policy constructed in Theorem 1, the corollary is evident.

The significance of Theorem 1 and its corollary is that for any optimization problem involving expected frequencies of state and decision in its cost criterion and constraints, only policies in C^M need be considered. That is, if R_1 (say) is any other policy that is optimal it can always be replaced by a Markovian policy R which is also optimal.

However, more can be said along these lines.

THEOREM 2. $H(i) = H^M(i) = \bar{H}^D(i) = \bar{H}^S(i)$.

Proof: First we prove that $H(i) \subset \bar{H}^D(i)$. Suppose the contrary; that is, there exists a point $X' = \{x'_{ja}(i)\} \in H(i)$ not contained in $\bar{H}^D(i)$. Since $\bar{H}^D(i)$ is a closed convex set there exists (Theorem 1 of Appendix C) a set of numbers $\{w'_{jk}\}$ such that

$$\sum_j \sum_a w'_{ja} x'_{ja} < \inf_{X \in \bar{H}^D(i)} \sum_j \sum_a w'_{ja} x_{ja}(i).$$

However, X' is a limit point of some policy $R \in C$. Hence,

$$\sum_j \sum_a w'_{ja} x'_{ja}(i) \geq \phi_R'(i) = \lim_{T \to \infty} \inf \sum_j \sum_a w'_{ja} x^R_{Tja}(i).$$

However, by Corollary 1 of Theorem 2, Chapter 3, there exists an $R^* \in C_D$ such that $\phi'_{R^*}(i) \leq \phi'_R(i)$. Thus we have a contradiction. We now prove that $\bar{H}^D(i) = \bar{H}^S(i)$. Clearly $\bar{H}^S(i) \supset \bar{H}^D(i)$ since $H^S(i) \supset H^D(i)$. However, $H^S(i) \subset H(i)$ and consequently, $\bar{H}^S(i) \subset \bar{H}(i)$, the closed convex hull of $H(i)$. But since $H(i) \subset \bar{H}^D(i)$ we must have $\bar{H}(i) \subset \bar{H}^D(i)$. Therefore $\bar{H}^S(i) \subset \bar{H}^D(i)$ and the equality is proved. The fact that $H(i) = H^M(i)$ follows from Theorem 1 and its corollary with $n = 1$; $H^M(i) \supset \bar{H}^D(i)$ follows from Theorem 1 applied to the policies $R_1, \ldots, R_n$ being selected from C_D. The theorem is proved.

As an application of Theorem 2 suppose $f(\cdot)$ is a continuous concave function defined over the closure of the points $X_T^R(i)$, $R \in C$, $T = 0, 1, \ldots$. With $Y_0 = i$, we wish to choose an $R \in C$ to minimize $\lim_{T \to \infty} \inf f(X_T^R(i))$.

Let $X^* \in H(i)$ be such that $f(X^*) = \min_{X \in H(i)} f(X)$. Since $H(i) = \bar{H}^D(i)$, and $f(X)$ is concave, it assumes its minimum (Theorem 2, Appendix B and Theorem 2a, Chapter 3) at an extreme point of $\bar{H}^D(i)$. However, every extreme point of $\bar{H}^D(i)$ is equal to $H^{R^*}(i)$ for some $R^* \in C_D$. Thus for some $R^* \in C_D$, $\lim_{T \to \infty} X_T^R(i) = X^{R^*}(i) = X^*$, and since f is continuous,

$$\lim_{T \to \infty} f(X_T^{R^*}(i)) = f(X^{R^*}(i))$$

$$= f(X^*)$$

$$= \min_{X \in H(i)} f(X)$$

$$= \min_{R \in C} \lim_{T \to \infty} \inf f(X_T^R(i)).$$

Therefore, f can be minimized by a policy $R^* \in C_D$.

The Main Theorems and Applications

THEOREM 3. If I has at most one ergodic class for every $R \in C_D$, then $H(i) = H^S(i)$.

Proof: In much the same way as Lemma 8 of Chapter 6 was proven we can show that if I has at most one ergodic class for every $R \in C_D$, this also holds for every $R \in C_S$. We shall show that under the hypothesis, $H^S(i)$ is closed and convex, in which case, $H^S(i) = \overline{H}^S(i)$ and the theorem follows using Theorem 2. For each $R \in C_S$, let $X = \{x_{ja}\} = \{\pi_j^R D_{ja}^R\}$ where π_j^R is the steady-state probability or long term expected frequency of state j under policy R. Let R^* and R^{**} be any two policies in C_S with X^* and X^{**} being their corresponding matrices. Let $X = \beta X^* + (1 - \beta) X^{**}$ where $0 \leq \beta \leq 1$ is arbitrary. In order to show that $H^S(i)$ is convex, we need to show that X corresponds to some $R \in C_S$. We note that X satisfies the system (the steady-state system of equations and inequalities)

$$x_{ja} \geq 0,$$

$$x_{ja} \geq 0,$$

$$\sum_i \sum_a x_{ia} q_{ij}(a) = \sum_a x_{ja} \geq 0, \quad j \in I,$$

$$\sum_i \sum_a x_{ia} = 1$$

since X^* and X^{**} satisfy the system. But we know that there corresponds a policy $R \in C_S$ yielding $X = \{\pi_j^R D_{ja}^R\}$; namely $D_{ja} = x_{ja}/\sum_a x_{ja}$, if $\sum_a x_{ja} > 0$, and D_{ja} arbitrary, if $\sum_a x_{ja} = 0$. Hence $H^S(i)$ is convex. To show that $H^S(i)$ is closed, let $\{R_v, v = 1, 2, \ldots\}$ be a sequence of policies in C_S such that $\{X_v, v = 1, 2, \ldots\}$, the sequence of corresponding X matrices, converges to X. We need to show that $X \in H^S(i)$. Since C_S is compact we can assume that $\{R_v, v = 1, 2, \ldots\}$ converges to R (say) $\in C_S$, for otherwise we can select a convergent subsequence of $\{R_v\}$ that does converge. However, the transition probabilities $\{p_{ij}\}$ are continuous functions of the policies in C_S and since the steady-state has a unique solution (Theorems 2 and 4, Appendix A), we must have that X corresponds to R. Thus $H^S(i)$ is closed as well as convex and the theorem is proved.

As in the application of Theorem 2, let $f(\cdot)$ be a continuous function over the closure of the possible points $X_T{}^R(i)$ for all $R \in C$, $T = 0$, $1, \ldots$, for a given $Y_0 = i$. Suppose the problem is to minimize $\lim_{T \to \infty} \inf f(X_T{}^R(i))$ over $R \in C$ subject to the constraint that $H^R(i) \subset G$, a given closed subset of $H(i)$. Since f is continuous and G is closed, an optimal policy will exist. Let R^* denote an optimal policy with $\lim_{T \to \infty} \inf f(X_T^{R^*}(i)) = f(X^*)$. Then by Theorem 3, there exists an $R^{**} \in C_S$ such that $X^{**} = X^*$, where X^{**} is the X matrix corresponding to R^{**}. That is, R^{**} is also optimal; consequently, in seeking an optimal policy, we need only consider those policies in class C_S.

Let us return to the problem described at the beginning of this chapter which is also Example 3 with $i = 0$ and $r = 1$. It was shown in Example 3 that $E\tau$, the expected recurrence time, and the probability under constraint are expressible as continuous functions of points in $H(i)$. Under reasonable conditions on the laws of motion the hypothesis of Theorem 3 will hold. Thus, it is possible in accordance with the above remark to restrict consideration to policies in C_S. However, for policies in C_S it is readily seen (Theorem 5 of Appendix A) that $E_R \tau = (\pi_0{}^R)^{-1}$ and $P_R\{Y_t = j,\ 1 \leq t \leq \tau \mid Y_0 = 0\} = \pi_j{}^R / \pi_0{}^R$. Thus, we can state the problem as that of minimizing $\pi_0{}^R$ subject to the constraint that $\pi_j{}^R \leq \pi_0{}^R \alpha$, where α is a given number $0 \leq \alpha \leq 1$. This problem can now be put into the linear programming form:

Minimize
$$\sum_a x_{0a}$$

subject to

$$x_{ia} \geq 0, \quad i \in I,$$

$$\sum_a x_{ia} = \sum_{la} x_{la} q_{li}(a), \quad i \in I,$$

The Main Theorems and Applications

$$\sum_i \sum_a x_{ia} = 1,$$

$$\sum_a x_{ia} \leq \alpha \sum_a x_{0a}.$$

Letting $D_{ia} = x_{ia}/\sum_a x_{ia}$ if $\sum_a x_{ia} > 0$ or D_{ia} arbitrary otherwise, yields the optimal policy $R \in C_S$.

In Theorem 4 of Chapter 3 we proved that $\sigma_R(i)$ is minimized by a policy $R \in C_D$. We now provide an alternative proof. Repeating the statement of the theorem:

THEOREM 4. Let j be the target state. If $P_R\{Y_t = j$ for some $t \geq 1 \mid Y_0 = i\} = 1$ for every $R \in C_D$, then there exists an $R \in C_D$ such that

$$\sigma_R(i) = \min_{R \in C} \sigma_R(i), \quad \text{for} \quad i \neq j.$$

Proof: Define $q_{ji}(a) = \beta$ when $i \neq j$, where $1 + 1/\beta$ is equal to the number of states in I. Since $w_{ja} \equiv 0$, we can define for every $R \in C$,

$$\sigma_R(j) = \beta \sum_{i \in I - \{j\}} \sigma_R(i).$$

(Note, in each case τ denotes $\min\{t \mid Y_t = j, t \geq 1\}$.) Clearly, if R minimizes $\sigma_R(j)$, it will also minimize $\sigma_R(i)$ for each $i \neq j$. We first argue that $E_R\{\tau \mid Y_0 = j\} < \infty$ for every $R \in C$. In Example 2 of this chapter, we have that $E_R\{\tau \mid Y_0 = j\} = \left(\lim_{T \to \infty} \sum_a x^R_{Tja}\right)^{-1}$, where R is any policy in the class of "renewal" policies. However, under the hypothesis of the theorem and the definition of $q_{ji}(a)$, the state space I is irreducible for every $R \in C_D$. Hence, by Theorem 3 if an R existed such that $E_R\{\tau \mid Y_0 = j\} = \infty$, then there must exist an $R \in C_S$ such that $E_R\{\tau \mid Y_0 = j\} = \infty$. But from Markov chain theory (Theorem 6 of Appendix A) this cannot be the case. Hence $E_R\{\tau \mid Y_0 = j\} < \infty$ for all R. Now from Example 2 we also have that $\sigma_R(j)$ is expressible as a continuous function of points in $H(i)$. In fact it can be shown that this function assumes its minimum at the extreme points of $H(i) = \overline{H}^D(i)$ (the equality of these two sets given by Theorem 2). Thus by the argument used in the application following Theorem 2, there exists a policy $R \in C_D$ that minimizes $\sigma_R(i)$. This proves the theorem.

State-Action Frequencies

The first three theorems of this chapter deal with expected state-action frequencies. In some applications it is desirable to have similar statements concerning the sample frequencies: that is, the actual frequencies of state-action combinations without taking expectations. If $R \in C_S$, the long-run frequencies and expected frequencies coincide with probability 1. However, if $R \notin C_S$ this may not be the case.
Let

$$Z_{tja} = 1, \quad \text{if} \quad Y_t = j, \quad A_t = a,$$
$$= 0, \quad \text{otherwise}.$$

Let

$$\bar{Z}_{Tja} = \frac{1}{T+1} \sum_{t=0}^{T} Z_{tja}$$

and $\bar{Z}_T$ denote the matrix of quantities $\{\bar{Z}_{Tja}\}$. For a fixed R, denote by ω a sample sequence of the joint process $\{Y_t, A_t, t = 0, 1, \ldots\}$. Let $U^R(\omega)$ be the set of limit points of $\{\bar{Z}_T{}^R, T = 0, 1, \ldots\}$. We have

THEOREM 5. *For each* $R \in C$, $P_R\{U^R(\omega) \subset \bar{H}\} = 1$, *where* $\bar{H}$ *is the closed convex hull of* $\bigcup_{j \in 1} H^D(j)$.

Before proceeding to the proof of Theorem 5, we shall need a preliminary inequality. Using the notation of Chapter 2, we set $V_T{}^*(i) = \min_{R \in C} E\left\{\sum_{t=0}^{T} W_t \mid Y_0 = i\right\}$. We state:

LEMMA 1. $\lim_{T \to \infty} \inf \min_{i \in I} V_T{}^*(i)/T \geq \min_{i \in I} \min_{R \in C} \phi_R(i)$.

State-Action Frequencies   99

Proof: If the inequality were not to hold, there would exist a sufficiently large T and state i, recurrent with respect to some policy such that

$$V_T^*(i) < T \min_{i \in I} \min_{R \in C} \phi_R(i),$$

from which one could construct a policy R with a $\phi_R(i)$ smaller than $\min_{i \in I} \min_{R \in C} \phi_R(i)$. We leave the details to the reader.

Proof of Theorem 5: Suppose the theorem is false. Let R be a policy such that $P_R\{U^R(\omega) \subset \bar{H}\} < 1$. Then there exists a sphere S with positive radius such that $S \cap \bar{H} = \varnothing$, the null set, and $P_R\{U^R(\omega) \cap S \neq \varnothing\} > 0$. This is so since $\bar{H}_C$, the complement of $\bar{H}$, can be covered by a denumerable number of such spheres S_v, $v = 1, 2, \ldots$ and

$$P_R\{U^R(\omega) \cap \bar{H}_C \neq \varnothing\} \leq \sum_{v=1}^{\infty} P_R\{U^R(\omega) \cap S_v\}.$$

Since $\bar{H}$ is a closed and bounded convex set and S is convex and the two sets are disjoint, the two sets can be separated by a hyper-plane; that is, by Theorem 1 of Appendix C there exists a set of numbers $\{w_{ia}\}$, $a \in K_i$, $i \in I$ such that $\sum_i \sum_a w_{ia} r_{ia} > \sum_i \sum_a w_{ia} s_{ia}$ for all $r = \{r_{ia}\} \in \bar{H}$ and $s = \{s_{ia}\} \in S$. Let $W_t = w_{ia}$ when $Y_t = i$, $A_t = a$ and note that

$$\frac{1}{T+1} \sum_{t=0}^{T} W_t = \sum_i \sum_a w_{ia} \bar{Z}_{Tia},$$

so that

$$\liminf_{T \to \infty} \frac{1}{T+1} \sum_{t=0}^{T} W_t = \sum_i \sum_a w_{ia} \bar{Z}_{ia}$$

for some point $\bar{Z} = \{\bar{Z}_{ia}\}$ in $U^R(\omega)$. We intend to show that the set of ω's such that $\liminf \frac{1}{T+1} \sum_{t=0}^{T} W_t < \min_{r \in H} \sum \sum w_{ia} r_{ia}$ has at most probability 0. If this is the case we must have $P_R\{U^R(\omega) \cap S \neq \varnothing\} = 0$, a contradiction proving the theorem. For a fixed N let

$$B_v = \sum_{t=(v-1)N+1}^{vN} W_t, \quad v = 1, \ldots, [T/N],$$

$$B' = \sum_{t=[T/N]N+1}^{T} W_t, \quad \text{if } [T/N] < T/N,$$

$$= 0, \quad \text{if } [T/N] = T/N,$$

where $[T/N]$ is the greatest integer less than or equal to T/N. Clearly $|B_v|$, $v = 1, \ldots, [T/N]$ and $|B'|$ are bounded and

$$E_R\{B_v | B_1, \ldots, B_{v-1}\}$$

$$\geq \min_{i \in I} \sum_j \sum_a w_{ja} \sum_{t=(v-1)N+1}^{vN} P_R\{Y_t = 0, A_t = a | Y_{(v-1)N} = i\}$$

$$= \min_{i \in I} \sum_j \sum_a w_{ja} \sum_{t=1}^{N} P_R\{Y_t = j, A_t = i | Y_0 = i\}$$

$$\geq \min_{i \in I} \{V_N^*(i) - \max_{a \in K_i} w_{ia}\}$$

$$\geq \min_{i \in I} V_N^*(i) - \max\{w_{ia}\}.$$

By Lemma 1, for any $\varepsilon > 0$, there exists an N such that

$$\frac{\max_{i,a} |w_{ia}|}{N} < \frac{\varepsilon}{2}$$

and

$$\frac{\min_{i \in I} V_N^*(i)}{N} \geq \min_{i \in I} \min_{R \in C} \phi_R(i) - \frac{\varepsilon}{2}$$

$$= \sum_i \sum_a w_{ia} r_{ia}^* - \frac{\varepsilon}{2},$$

where $r^* = \{r_{ia}^*\}$ is such that $\sum_i \sum_a w_{ia}^* r_{ia} = \min_{i \in I} \min_{R \in C} \phi_R(i)$. Thus, for the value of N and for $v = 1, \ldots, [T/N]$ (the greatest integer less than or equal to T/N), we have

$$E_R\{B_v | B_1, \ldots, B_{v-1}\} \geq N \sum_i \sum_a w_{ia} r_{ia}^* - N\varepsilon.$$

However, by Theorem 5 of Appendix B, we have

$$\lim_{T \to \infty} [T/N]^{-1} \sum_{v=1}^{[T/N]} \{B_v - E(B_v \mid B_1, \ldots, B_{v-1})\} = 0$$

with probability 1. Consequently,

$$\liminf_{T \to \infty} [T/N]^{-1} \sum_{v=1}^{[T/N]} B_v \geq N \left(\sum_i \sum_a w_{ia} r^*_{ia} - \varepsilon \right)$$

with probability 1. But then, with probability 1,

$$\liminf_{T \to \infty} \frac{1}{T+1} \sum_{t=0}^{T} W_t$$

$$= \liminf_{T \to \infty} \frac{1}{T+1} \left\{ W_0 + \sum_{v=1}^{[T/N]} B_v + B' \right\}$$

$$\geq \liminf_{T \to \infty} \frac{1}{T+1} \sum_{v=1}^{[T/N]} B_v + \liminf_{T \to \infty} \frac{1}{T+1} (B' + W_0)$$

$$= \liminf_{T \to \infty} \frac{1}{T+1} \sum_{v=1}^{[T/N]} B_v$$

$$= N^{-1} \liminf_{T \to \infty} [T/N]^{-1} \sum_{v=1}^{[T/N]} B_v$$

$$\geq \sum_i \sum_a w_{ia} r^*_{ia} - \varepsilon.$$

Since ε is arbitrary, we have

$$\liminf_{T \to \infty} \frac{1}{T+1} \sum_{t=0}^{T} W_t \geq \sum_i \sum_a w_{ia} r^*_{ia}$$

with probability 1, and the theorem is proved.

As an application, suppose $f(\cdot)$ is a continuous function defined over the closure of the possible values of $\bar{Z}_T$, $T = 0, 1, \ldots$ and $\bar{H}$, and it is of interest to find $R \in C$ which minimizes $E \liminf_{T \to \infty} f(\bar{Z}_T)$ (as distinct from minimizing $\liminf_{T \to \infty} f(X_T) = \liminf_{T \to \infty} f(E\bar{Z}_T)$). Assume that, for each $R \in C_D$, I is irreducible. Then $\bar{H} = H^S(i) = H^S$ independent of

i by Theorem 3 and Theorem 4 of Appendix A. Since $f(\cdot)$ is continuous, then $\lim_{T\to\infty} \inf f(\bar{Z}_T) \geq \min_{r \in \bar{H}} f(r) = f(r_0)$ with probability 1. Hence,

$$E \lim \inf f(\bar{Z}_T) \geq \min f(r_0).$$

Since $\bar{H} = H^S$, there exists a policy $R \in C_S$ such that $\lim_{T\to\infty} Z_T = r_0$ with probability 1. Therefore, under this policy, $E_R \lim_{T\to\infty} \inf f(\bar{Z}_T) = f(r_0)$.

Bibliographical Remarks

Theorem 1 is a slightly more general form of a result obtained by Derman and Strauch [25]. The general form was given by Strauch and Veinott [50], from which follows the equality of $H(i)$ with $H^M(i)$ in Theorem 2. The remaining results of this chapter are due to the author [16, 18].

8

Optimal Stopping of a Markov Chain

Statement of the Problem

Let us suppose $\{Y_t, t = 0, 1, \ldots\}$ is a finite state Markov chain with stationary transition probabilities $\{p_{ij}\}$. Let us suppose there exists an absorbing state 0 (that is, $p_{00} = 1$) in the state space I such that $P\{Y_t = 0 \text{ for some } t \geq 1 \mid Y_0 = i\} = 1$ for every $i \in I$. Let $\{w_i, i \in I\}$ denote nonnegative numerical values associated with each state. When the chain is absorbed at state 0, we can think of the process as having been stopped at that point in time and we receive the value w_0 associated with the state 0. However, we can also think of stopping the process at any point in time prior to absorption and receiving the value w_i if i is the state of the chain when the process is stopped. If our aim is to receive the highest possible numerical value and if $w_0 < \max_{i \in I}\{w_i\}$, then clearly we would not necessarily wait for absorption before stopping the process.

By a stopping time τ, we mean a rule that prescribes the time to stop the process; $\tau = t$ means that the process is stopped at time t and the information for stopping the process at time t must be confined to the values of the variables $Y_0, \ldots, Y_t$. We shall assume for all stopping times τ considered that if $\tau_1 = \min\{t \mid Y_t = 0, t \geq 1\}$, then $\tau \leq \tau_1$.

By a *stopped process* $\{\tilde{Y}_t, t = 0, 1, \ldots\}$ *determined by a stopping time* τ we mean

$$\tilde{Y}_t = Y_t, \quad t \leq \tau,$$
$$= Y_\tau, \quad \text{if} \quad t > \tau.$$

The problem of this chapter is to determine the stopping time τ such that $E\{w_{Y_\tau} \mid Y_0 = i\}, i \in I - \{0\}$, is maximized.

Stopping Problem as an Expected Average Gain Problem

We first remark that an optimal stopping time does exist and in fact is of the form that the prescription as to when to stop the process need only be a function of the state of the process at the time of stopping; that is, I will be dichotimized into states where the process is stopped and states where the process is not stopped. To see this we need only to observe that the problem can be reformulated so as to be of the form discussed in Chapter 6. At each state there are two possible actions. Action 1 continues the process according to the transition probabilities $\{p_{ij}\}$; action 2 at state i transforms i into an absorbing state. At state 0 the two actions coincide. That is,

$$q_{ij}(1) = p_{ij}, \quad q_{ij}(2) = \delta_{ij}, \quad i \in I, \quad j \in I.$$

Set

$$w_{i1} = 0, \quad w_{i2} = w_i, \quad i \in I - \{0\},$$

and
$$w_{01} = w_{02} = w_0.$$

Consider the problem of maximizing $\phi_R(i)$, the expected average cost per unit time, over all possible policies in C. Notice that the class of stopping times is a subclass of the class C of all policies. This is the subclass of policies such that whenever action 2 is dictated at a state i and time $t = \tau$, then action 2 is dictated for all $t > \tau$. Notice also, that for such a policy $R = \tau$ (say), $\phi_R(i) = E\{w_{Y_\tau} | Y_0 = i\}$. Thus $\max_{R \in C} \phi_R(i) \geq \max_\tau E\{w_{Y_\tau} | Y_0 = i\}$. However, by Theorem 2, Chapter 3, or its Corollary 1, $\phi_R(i)$ is maximized for each $i \in I$ by a policy $R \in C_D$. But each $R \in C_D$ is a stopping time since, if action 2 is prescribed at state i, the process will remain in state i and continue to prescribe action 2. Thus,

$$\max_{R \in C} \phi_R(i) = \max_{R \in C_D} \phi_R(i)$$
$$= \max_\tau E\{w_{Y_\tau} | Y_0 = i\}, \quad i \in I,$$

where the optimal stopping time τ is the policy $R \in C_D$ that maximizes $\phi_R(i)$, $i \in I$.

Of course, it follows that the computational methods of Chapter 6 can be used to obtain an optimal stopping time.

A Different Approach

We return to the original problem formulation of this chapter and offer another approach. Let

$$M(i) = \max_\tau E\{w_{Y_\tau} | Y_0 = i\}, \quad i \in I.$$

By the remarks of the previous section, the optimal stopping time τ need only be a time invariant function of the state of the process, so

that we have the dynamic programming functional equations
$$M(0) = w_0$$
$$M(i) = \max\left\{w_i, \sum_j p_{ij} M(j)\right\}, \quad i \in I - \{0\}; \tag{1}$$

thus, the optimal stopping time takes the form of stopping the process at those values of i where $w_i \geq \sum_j p_{ij} M(j)$, $i \in I$. If $M(i)$, $i \in I$, were a known function, the optimal stopping time would be known. The following discussion is intended to provide methods for determining $M(i)$, $i \in I$.

By a *super-regular function $f(i)$, $i \in I$, with respect to $\{p_{ij}\}$*, we mean a nonnegative function satisfying
$$\sum_j p_{ij} f(j) \leq f(i), \quad i \in I. \tag{2}$$

We first prove:

LEMMA 1. *Let τ be any stopping time. If $f(i)$, $i \in I$, is any function such that $f(i) \geq w_i$, $i \in I$, then*
$$E\{w_{Y_\tau} | Y_0 = i\} \leq E\{f(Y_\tau) | Y_0 = i\}, \quad i \in I.$$

Proof:
$$E\{f(Y_\tau) | Y_0 = i\} = \sum_j E\{f(Y_\tau) | Y_0 = i, \ Y_\tau = j\} P\{Y_\tau = j | Y_0 = i\}$$
$$= \sum_j f(j) P\{Y_\tau = j | Y_0 = i\}$$
$$\geq \sum_j w_j P\{Y_\tau = j | Y_0 = i\}$$
$$= \sum_j E\{w_{Y_\tau} | Y_0 = i, \ Y_\tau = j\} P\{Y_\tau = j | Y_0 = i\}$$
$$= E\{w_{Y_\tau} | Y_0 = i\}.$$

LEMMA 2. *Let τ be any stopping time. If $f(i)$, $i \in I$, is super-regular, then*
$$E\{f(Y_\tau) | Y_0 = i\} \leq f(i), \quad i \in I.$$

A Different Approach

Proof: Let Ω be the space of all sequences $\omega = \{i_k, k = 0, 1, \ldots\}$, where the range of each coordinate of ω is I. Because $\{Y_t\}$ is eventually stopped or absorbed at 0, all the probability mass on Ω is concentrated on a denumerable subset of Ω. In what follows, $P\{\omega\}$ is to be interpreted as $P\{Y_k = i_k, k = 0, 1, \ldots\}$. Let E_n denote any subset of Ω determined by conditions on $i_0, i_1, \ldots, i_n$ (that is, on the first $n+1$ coordinates of ω). Since f is super-regular, we have

$$\sum_{\omega \in E_n} P\{\omega \mid Y_0 = i\} f(Y_{n+1}(\omega))$$

$$= \sum_{\omega \in E_n} \sum_j \sum_l P\{\omega, Y_n = l, Y_{n+1} = j \mid Y_0 = i\} f(j)$$

$$= \sum_{\omega \in E_n} \sum_j \sum_l P\{\omega, Y_n = l \mid Y_0 = i\} p_{lj} f(j)$$

$$\leq \sum_{\omega \in E_n} \sum_l P\{\omega, Y_n = l \mid Y_0 = i\} f(l)$$

$$= \sum_{\omega \in E_n} P(\omega \mid Y_0 = i) f(Y_n(\omega)).$$

Recall that $\{\tilde{Y}_n, n = 0, 1, \ldots\}$ denotes the stopped process. We now show that for each $n = 0, 1, \ldots$,

$$E\{f(\tilde{Y}_n) \mid Y_0 = i\} \leq f(i), \quad i \in I. \tag{3}$$

Noting that $\{\omega \mid \tau \leq n\}$ and $\{\omega \mid \tau > n\}$ are both subsets of Ω of the form E_n, we can write using (3)

$$E\{f(\tilde{Y}_{n+1}) \mid Y_0 = i\}$$

$$= \sum_{\{\omega \mid \tau > n\}} f(\tilde{Y}_{n+1}(\omega)) P\{\omega \mid Y_0 = i\} + \sum_{\{\omega \mid \tau \leq n\}} f(\tilde{Y}_{n+1}(\omega)) P\{\omega \mid Y_0 = i\}$$

$$= \sum_{\{\omega \mid \tau > n\}} f(Y_{n+1}(\omega)) P\{\omega \mid Y_0 = i\} + \sum_{\{\omega \mid \tau \leq n\}} f(\tilde{Y}_n(\omega)) P\{\omega \mid Y_0 = i\}$$

$$\leq \sum_{\{\omega \mid \tau > n\}} f(Y_n(\omega)) P\{\omega \mid Y_0 = i\} + \sum_{\{\omega \mid \tau \leq n\}} f(\tilde{Y}_n(\omega)) P\{\omega \mid Y_0 = i\}$$

$$= \sum_{\{\omega \mid \tau > n\}} f(\tilde{Y}_n(\omega)) P\{\omega \mid Y_0 = i\} + \sum_{\{\omega \mid \tau \leq n\}} f(\tilde{Y}_n(\omega)) P\{\omega \mid Y_0 = i\}$$

$$= E\{f(\tilde{Y}_n) \mid Y_0 = i\}.$$

Since $E\{f(\tilde{Y}_0) \mid Y_0 = i\} = f(i)$, Eq. (4) holds on iterating the above inequality. Since $\tau < \infty$ with probability 1 (because τ is less than or equal to the time of absorption at state 0), we have that $\lim_{n \to \infty} Y_n(\omega) = Y_\tau(\omega)$ with probability 1.

Since interchange of limit and expectation are valid here, we have

$$E\{f(\tilde{Y}_\tau) | Y_0 = i\} = \lim_{n \to \infty} E\{f(\tilde{Y}_n) | Y_0 = i\}$$

$$\leq f(i), \quad i \in I.$$

and the lemma is proved.

We now define the *smallest super-regular function dominating* $\{w_i, i \in I\}$ as that function $\{s(i), i \in I\}$ satisfying the conditions that

(i) s is super-regular,
(ii) $s(i) \geq w_i, i \in I$,
(iii) $s(i) \leq f(i), i \in I$, whenever f is super-regular and $f(i) \geq w_i, i \in I$.

If f and g are two super-regular functions then $h = \min(f, g)$ is also super-regular since

$$\sum_j p_{ij} \min(f(j), g(j)) \leq \min\left(\sum_j p_{ij} f(j), \sum_j p_{ij} g(j)\right)$$

$$\leq \min(f(j), g(j)), \quad i \in I.$$

Thus, s can be defined as the lower envelope of all superregular functions dominating $\{w_i, i \in I\}$. Clearly, one and only one such function exists.

The main result relating $M(i)$ to the notion of the smallest super-regular function dominating $\{w_i, i \in I\}$ is

THEOREM 1. The function $\{M(i), i \in I\}$ of (1) is equivalent to the smallest super-regular function dominating $\{w_i, i \in I\}$.

Proof: Clearly $M(i)$ satisfies condition (ii). Also, from Eq. (1), if $M(i) = w_i$, then $M(i) \geq \sum_j p_{ij} M(j)$; otherwise, $M(i) = \sum_j p_{ij} M(j)$. Hence, condition (i) holds. To show that condition (iii) holds, suppose f is super-regular and $f(i) \geq w_i, i \in I$. Let τ be the optimal stopping time.

A Different Approach

Then for each $i \in I$, by Lemma 1 and 2,

$$M(i) = E\{w_{Y_\tau} \mid Y_0 = i\}$$
$$\leq E\{f(Y_\tau) \mid Y_0 = i\}$$
$$\leq f(i).$$

Thus, condition (iii) holds and the theorem is proved.

We now can determine $\{M(i), i \in I\}$ by solving a linear programming problem as given in:

THEOREM 2. If $\{v_i^*, i \in I\}$ is an optimal solution to the linear programming problem to

minimize
$$\sum_{i \in I} v_i$$

subject to
$$\sum_j p_{ij} v_j \leq v_i,$$
$$v_i \geq w_i, \quad i \in I,$$

then $M(i) = v_i^*, i \in I.$

Proof: From the constraints of linear programming problem, the function $\{v_i^*, i \in I\}$ is super-regular and dominates $\{w_i, i \in I\}$. By Theorem 2, $s(i) = M(i), i \in I$. If $\{M(i), i \in I\}$ is not equal to $\{v_i^*, i \in I\}$ then $M(i) \leq v_i^*, i \in I$, with strict inequality holding for at least one i. However, then $\{v_i^*, i \in I\}$ cannot be the optimal solution to the linear programming problem, a contradiction proving the theorem.

Another method for calculating $\{M(i), i \in I\}$ is a method of successive approximations. Let $\{f_0(i), i \in I\}$ be a given function. Define recursively, $f_n(i) = \max\{f_0(i), \sum_j p_{ij} f_{n-1}(j)\}$, $i \in I$ for $n = 1, 2, \ldots$. We have:

THEOREM 3. If $f_0(i) = w_i$, $i \in I$, then $M(i) = \lim_{n \to \infty} f_r(i)$, $i \in I$.

Proof: It is easily established that $f_n(i) = E\{w_{Y_{\tau_n}} | Y_0 = i\}$, $i \in I$, where τ_n is the optimal stopping time among the class of all stopping numbers such that $\tau \leq n$ with probability 1. As $\{f_n\}$ is a nondecreasing sequence with $f_n(i) \leq M(i)$, $i \in I$, then $f(i) = \lim_{n \to \infty} f_n(i) \leq M(i)$. We also have

$$f(i) = \max\{w_i, \sum_j p_{ij} f(j)\}, \qquad i \in I,$$

and clearly f is a super-regular function that dominates $\{w_i, i \in I\}$. However, by Theorem 2, we must have that $f(i) = M(i)$, $i \in I$, so that the theorem is proved.

A third method for calculating $\{M(i), i \in I\}$ is to solve the system (1). That this is true follows from:

THEOREM 4. If $\{f(i), i \in I\}$ satisfies

$$f(0) = w_0$$

$$f(i) = \max\{w_i, \sum_j p_{ij} f(j)\}, \qquad i \in I - \{0\},$$

then $f(i) = M(i)$, $i \in I$.

Proof: If $\{f(i), i \in I\}$ satisfies (1) then it is super-regular with $f(i) \geq w_i$, $i \in I$. Since $M(i)$ is the unique smallest super-regular function dominating $\{w_i, i \in I\}$, then

$$\Delta_i = f(i) - M(i)$$
$$\geq 0, \qquad i \in I.$$

A Different Approach

On subtracting $M(i)$ from $f(i)$ in (1), we obtain

$$\Delta_i \leq \sum_j p_{ij} \Delta_j, \qquad i \in I.$$

However, on iterating, we obtain that

$$\Delta_i \leq \sum_j p_{ij}^{(t)} \Delta_j, \qquad i \in I,$$

and since $j = 0$ is an absorbing state with all other states being transient, $\lim_{t \to 0} p_{ij}^{(t)} = 0$ for $j \neq 0$. Thus

$$\Delta_i \leq \Delta_0$$
$$= 0, \qquad i \in I,$$

which proves the theorem.

It is sometimes possible, without knowing $\{M(i), i \in I\}$, to determine that a state i as one at which the process is stopped under an optimal stopping time. More explicitly we state:

THEOREM 5. Let $\{f(j), j \in I\}$ be any super-regular function that dominates $\{w_j, j \in I\}$ (that is, $f(j) \geq w_j, j \in I$). If for some i, $\sum_j p_{ij} f(j) = w_i$, then $M(i) = w_i$; that is, i is a state where the process is stopped under an optimal policy.

Proof: Since $w_i \leq M(i) \leq f(i)$,

$$M(i) = \max\left\{w_i, \sum_j p_{ij} M(j)\right\}$$
$$\leq \max\left\{w_i, \sum_j p_{ij} f(j)\right\}$$
$$= w_i;$$

hence, equality must hold.

See Problem 3 for an application of Theorem 5.

Computational Example

Suppose $I = 0, 1, 2$, where

$$\begin{pmatrix} p_{00} & p_{01} & p_{02} \\ p_{10} & p_{11} & p_{12} \\ p_{20} & p_{21} & p_{22} \end{pmatrix} = \begin{pmatrix} 1 & 0 & 0 \\ \frac{1}{3} & \frac{1}{3} & \frac{1}{3} \\ \frac{1}{2} & \frac{1}{2} & \frac{1}{2} \end{pmatrix},$$

and $(w_0, w_1, w_2) = (0, 2, 1)$. We want to find τ to minimize Ew_{Y_τ}. Let us solve for $M(i)$, $i = 0, 1, 2$, by linear programming. Since we know that $M(0) = 0$, the linear programming problem can be stated as finding those variables v_i, v_2

to minimize

$$v_1 + v_2$$

subject to

$$-\tfrac{2}{3}v_1 + \tfrac{1}{3}v_2 \leqq 0$$
$$-\tfrac{3}{4}v_1 + \tfrac{1}{4}v_2 \leqq 0$$

and

$$v_1 \geqq 2, \qquad v_2 \geqq 1.$$

Solving, we find $v_i^* = 2$, $v_i^* = 1$ as an optimal solution. Thus, $M(0) = 0$, $M(1) = 2$, $M(2) = 1$, where

$$w_1 = 2$$
$$> \tfrac{1}{3}(2 + 1)$$

and

$$w_2 = 1$$
$$< \tfrac{1}{4}(2 + 1).$$

Thus, the optimal stopping time is always $\tau = 0$.

The Dual Linear Programming Problem

Let us consider the dual linear programming problem for obtaining $\{M(i), i \in I\}$. First, it is convenient to rewrite the primal problem:

Minimize
$$\sum_i \beta_i u_i$$
subject to
$$u_i \geq 0, \quad i \in I,$$
$$\sum_j (\delta_{ij} - p_{ij}) u_j \geq -\sum_j (\delta_{ij} - p_{ij}) w_j$$
$$= \sum_j p_{ij} w_j - w_i$$
$$= \gamma_i, \quad i \in I \quad \text{(say)}.$$

This problem was obtained by setting $u_i = v_i - w_i$ in the original primal problem; the constant term in the objective function has been dropped. The dual problem is:

Maximize
$$\sum_i \gamma_i x_i$$
subject to
$$x_i \geq 0, \quad i \in I,$$
$$\sum_i x_i(\delta_{ij} - p_{ij}) \leq \beta_j, \quad j \in I.$$

From Theorem 2 of Appendix A, one can argue that for every possible stopping set S, the values $\{\tilde{x}_i\}$, equal to the expected number of entries into state i from time $t = 0$ up to but not including the time of entry into S, where $P(Y_0 = i) = \beta_i, i \in I$, are feasible solutions to the dual problem. However, the objective function under one of these solutions for τ given by the stopping set S, can be seen to be $Ew_{Y_\tau} - \sum \beta_i w_i$. Thus if S is the optimal stopping set, the objective function must equal $\sum_i \beta_i u_i$, where $\{u_i\}$ is the optimal solution to the primal problem. By the duality theorem (Theorem 4 of Appendix C), at least one optimal solution of the dual problem must be $\{\tilde{x}_i\}$, where S is the optimal stopping set. In any case, using Theorem 4 of Appendix C, part of the optimal stopping set can easily be extracted from the dual problem solution; namely, set $u_j = 0$ when the jth constraint in the dual problem solution holds with strict inequality prevailing. However, $u_j = 0$ implies that $j \in S$.

Some Other Forms of the Stopping Problem

The problem formulation of this chapter includes the case where a cost is incurred for each period that the process continues. The cost can be a function of the state of the process; that is, there is a cost c_i each time the process is in state i, $i \in I$. The problem is to maximize

$$E\left\{w_{Y_\tau} - \sum_{t=0}^{\tau} c_{Y_t} \mid Y_0 = i\right\}, \qquad i \in I,$$

by selecting the best stopping number τ.

Let

$$C(i) = E\left\{\sum_{t=0}^{\tau_1} c_{Y_t} \mid Y_0 = i\right\}, \qquad i \in I,$$

where τ_1 denotes the stopping time that waits until $Y_t = 0$ for the first time; that is, $\tau_1 = \min\{t \mid Y_t = 0, t \geq 1\}$. For any stopping time τ, let us note that

$$E\left\{\sum_{t=\tau+1}^{\tau_1} c_{Y_t} \mid Y_0 = i\right\}$$

$$= \sum_j \sum_{n=0}^{\infty} E\left\{\sum_{t=\tau+1}^{\tau_1} c_{Y_t} \mid Y_\tau = j, \ Y_0 = i, \ \tau = n\right\}$$
$$\times P\{Y_\tau = j, \ \tau = n \mid Y_0 = i\}$$
$$= \sum_j \sum_{n=0}^{\infty} C(j) P\{Y_\tau = j, \ \tau = n \mid Y_0 = i\}$$
$$= \sum_j C(j) P\{Y_\tau = j \mid Y_0 = i\}, \qquad i \in I.$$

Some Other Forms of the Stopping Problem

Then, for any stopping time τ,

$$E\left\{w_{Y_\tau} - \sum_{t=0}^{\tau} c_{Y_t} \mid Y_0 = i\right\}$$

$$= E\{w_{Y_\tau} \mid Y_0 = i\} + E\left\{\sum_{t=\tau+1}^{\tau_1} c_{Y_t} \mid Y_0 = i\right\} - C(i)$$

$$= \sum_j w_j P\{Y_\tau = j \mid Y_0 = i\} + \sum_j C(j)P\{Y_\tau = i\} - C(i)$$

$$= \sum_j (w_j + C(j))P\{Y_\tau = j \mid Y_0 = i\} - C(i)$$

$$= E\{w_{Y_\tau} + C(Y_\tau) \mid Y_0 = i\} - C(i).$$

Therefore, on letting $w_i' = w_i + C(i)$, $i \in I$, and solving the original stopping problem with respect to the values $\{w_i', i \in I\}$, we will obtain an optimal stopping time for the problem with costs.

The problem formulation also includes the problem of finding a stopping time τ to maximize

$$E\{\alpha^\tau w_{Y_\tau} \mid Y_0 = i\}, \qquad i \in I,$$

where α is a number between 0 and 1. It is well to note that for this problem, it is not necessary to have an absorbing state in order to have a nontrivial problem. The presence of the factor α^τ makes it imperative that we do not wait too long before stopping the process. However, we now show that by introducing an additional state, the problem can be reverted to its original form.

Let us consider a related Markov chain $\{Y_t', t = 0, 1, \ldots\}$ over the state space $I' = I \cup \{0\}$ where 0 is an absorbing state of $\{Y_t', t = 0, 1, \ldots\}$. More specifically, we let $p_{00}' = 1$; $p_{i0}' = 1 - \alpha$, $i \in I$; $p_{ij}' = \alpha p_{ij}$, $i, j \in I$; $w_0' = 0$, $w_i' = w_i$, $i \in I$. For any stopping time τ' to stop the process, $\{Y_t', t = 0, 1, \ldots\}$, which is a function only of the state of the process, relate the stopping time τ to stop $\{Y_t, t = 0, 1, \ldots\}$ by defining τ to stop $\{Y_t\}$ at those states $i \neq 0$ at which τ' stops $\{Y_t'\}$.

However, for any stopping time τ', we have

$$E\{w'_{Y'_{\tau'}} | Y_0' = i\}$$

$$= \sum_{j \in I} \sum_{t=0}^{\infty} w_j' P\{Y_t' = j, \quad \tau' = t | Y_0 = i\}$$

$$= \sum_{j \in I} \sum_{t=0}^{\infty} w_j' P\{Y_t' = j, \quad Y_n' \neq 0, \quad 1 \leq n < t, \quad \tau' = t | Y_0 = i\}$$

$$= \sum_{j \in I} \sum_{t=0}^{\infty} w_j \alpha^t P\{Y_t = j, \quad Y_n \neq 0, \quad 1 \leq n < t, \quad \tau = t | Y_0 = i\}$$

$$= E\{\alpha^\tau w_{Y_\tau} | Y_0 = i\}.$$

Thus, if τ' is optimal for stopping $\{Y_t', t = 0, 1, \ldots\}$ in order to maximize $E\{w'_{Y'_\tau} | Y_0' = i\}$, $i \in I$, then τ is optimal for maximizing

$$E\{\alpha^\tau w_{Y_\tau} | Y_0 = i\}, i \in I.$$

We point out that just as the maximization of $E\{w_{Y_\tau} | Y_0 = i\}$ can be viewed as an expected average gain maximization, the maximization of $E\{\alpha^\tau w_{Y_\tau} | Y_0 = i\}$ can be viewed as an expected discounted gain maximization. The constant α being the discount factor. We leave the details of establishing the equivalence to the reader. Of course, then, the computational methods of Chapter 4 are applicable.

Bibliographical Remarks

A substantial literature exists with respect to the problem of when to stop a stochastic process. Much of the literature involves the methods of martingales. Papers by Snell [49], Derman and Sachs [24], and Chow and Robbins [10] are early works along these lines. The methods presented in this chapter are due to Dynkin [29]. See Breiman [8].

The proof of Lemma 2 is one encountered in the study of martingales (see Doob [28], p. 300); it differs from the proof given by Dynkin.

Taylor [51, 52] has exploited the Dynkin approach in connection with continuous time and space stopping problems.

The optimal policy of Problem 4 follows from a general theorem proved by Derman and Sachs [24] and also by Chow and Robbins [10]. Breiman [8] refers to the set of conditions as the absolutely monotone case. The result with proof of Problem 5 also appears in Breiman [8]. He refers to this as the monotone case.

Our modified primal problem in the discussion of the dual linear programming problem is obtained by Breiman [8] by other means. That which we denote by γ_i is the negative of that which Breiman calls entrance fees for our stopping problem. These entrance fees also appear in Problems 4 and 5.

Problems

(1) Suppose $I = \{0, 1, 2, 3\}$, $(w_0, w_1, w_2, w_3) = (0, 1, 2, 1)$,

$$\begin{pmatrix} p_{00} & p_{01} & p_{02} & p_{03} \\ p_{10} & p_{11} & p_{12} & p_{13} \\ p_{20} & p_{21} & p_{22} & p_{23} \\ p_{30} & p_{31} & p_{32} & p_{33} \end{pmatrix} = \begin{pmatrix} 1 & 0 & 0 & 0 \\ \frac{1}{8} & \frac{1}{8} & \frac{1}{2} & \frac{1}{4} \\ \frac{1}{3} & \frac{1}{3} & \frac{1}{3} & 0 \\ \frac{1}{4} & \frac{1}{8} & \frac{1}{2} & \frac{1}{8} \end{pmatrix}$$

Find τ that maximizes $E\{w_{Y_\tau}\}$.

(2) If $f(i), i \in I$, is such that $\sum_j p_{ij} f(j) = f(i), i \in I$, then show that

$$E\{f(Y_\tau) \mid Y_0 = i\} = f(i), \quad i \in I,$$

for any stopping time τ.

(3) Suppose there are n objects with associated distinct values $v_1, v_2, \ldots, v_n$. We define the following selection process: An object is selected at random. If its value is acceptable, then the process of selection terminates with the value of the selected object given to the decision maker. If the value of the object is unacceptable, then the object is discarded and another random selection is made from the remaining $n - 1$ objects.

The selection process continues in this manner until an object is accepted. If all n objects have been rejected then the value received is zero. Determine a stopping procedure that maximizes the probability of choice of the most valuable object. Consider only stopping procedures that do not accept an object whose value is less than the value of one already rejected.

Solution (Dynkin [29]): Consider the state space $I = \{1, 2, \ldots, n, 0\}$, with $Y_1 = 1$. Let $Y_2 = i$ if the ith object selected is the first to have a value greater than the first selected; $Y_3 = j$ if the jth object selected is the first object to have a value greater than that associated with the value of the ith object selected. In general, Y_t is the number of the object selected which has value exceeding the value of the (Y_{t-1})th object selected; $Y_t = 0$ when n objects have been rejected. Clearly, $p_{ij} = 0$ if $1 \leq j \leq i$. For $i < j$,

$$p_{ij} = \frac{i}{j(j-1)}, \qquad p_{i0} = 1 - \sum_{j=i+1}^{n} p_{ij}.$$

Also, whenever $Y_t = i$ and the object is accepted, the probability that the object is most valuable is i/n. Therefore, we set $w_i = i/n$, $i = 1, \ldots, n$. Let i^* be the largest integer for which $\sum_{j=i^*}^{n-1} i/j > 1$. One can verify that the function $f(j) = \max(i^*/n, j/n)$ is a superregular function which dominates $\{w_j, j \in I\}$. Also, $\sum_j p_{ij} f(j) = w_i$ for $i \geq i^*$. Thus, by Theorem 5, states i for which $i \geq i^*$ are states at which an optimal stopping time τ stops and $M(i) = i/n$ for $i \geq i^*$. Since $p_{ij} = 0$ for $j \leq i$,

$$M(i^* - 1) = \left\{\frac{i^* - 1}{n}, \sum_{j=i}^{n} \frac{i^* - 1}{j(j-1)} j/n\right\} > \frac{i^* - 1}{n};$$

hence, at state $i^* - 1$ an optimal stopping time does not stop. Repeating the argument successively, the same holds for state $i^* - 2, \ldots, 1$. Summarizing, τ stops at states $i^*, \ldots, n$ and does not stop at states $i, \ldots, i^* - 1$.

(4) Suppose E is a set of states such that $p_{ij} = 0$ for every $i \in E$ and $j \notin E$, $\sum_j p_{ij} w_j \geq w_i$ for every $i \notin E$, and $\sum_j p_{ij} w_j \leq w_i$ for every $i \in E$. Then show that an optimal policy stops for all $i \in E$, and continues for all $i \notin E$.

(5) Suppose $f(i) = w_i - \sum_j p_{ij} w_j$, $i \in I$, is a nonincreasing function and $\sum_j p_{ij} g(j)$, $i \in I$, is nonincreasing whenever $g(i), i \in I$, is nonincreasing. Show that the optimal policy is of the form: stop for all $i \geq i^*$ and continue for all $i < i^*$ where i^* is a state that must be determined.

Proof (Breiman [8]): Let

$$H(i) = M(i) - w_i$$
$$= \max\left\{0, \sum_j p_{ij} M(j) - w_i\right\}$$
$$= \max\left\{0, \sum_j p_{ij} H(j) - f(i)\right\}, \quad i \in I.$$

Using Theorem 3, show that $H(i)$ is nonincreasing from which it will follow that $H(i) = 0$ for all $i \geq i^*$ and $H(i) < 0$ for $i < i^*$.

9

Some Applications

1 A Replacement Model

A common activity is the periodic inspection of some system, or one of its components, as part of a procedure for keeping it operative. After each inspection, an action must be taken as to whether or not to alter the system at that time. The problem is that of determining, according to some appropriate cost criterion, the optimal policy for taking actions.

More specifically, suppose a unit (a system, a component of a system, a piece of operating equipment, etc.) is inspected at equally spaced points in time and that after each inspection it is classified into one of $L+1$ states $0, 1, \ldots, L$. Then $\{Y_t\}$ is the sequence of states. A unit is in state 0 if and only if it is new; a unit is in state L if and only if it is inoperative. We assume that at states $1, \ldots, L-1$, there are two possible actions: $a = 1$ is not to replace the unit, $a = 2$ is to replace the unit. At state 0 only one action is possible, not to replace. At state L only one action is possible, to replace. Accordingly, we set $q_{ij}(1) = p_{ij}$, $i = 0, \ldots, L, j = 0, \ldots, L$ with $P_{i0} = 0, i = 0, \ldots, L,$ and $p_{L0} = 1$; $q_{i0}(2) = 1, i = 1, \ldots, L-1$. We assume the $\{p_{ij}\}$ are such that for every i $(i = 0, \ldots, L-1) p_{iL}^{(t)} > 0$ for some $t \geq 1$. This implies that a unit not replaced will eventually become inoperative with probability 1.

We assume two types of cost, the cost to replace an operative unit and the cost to replace an inoperative unit. That is, we set

$$w_{i1} = 0, \qquad i = 0, \ldots, L-1,$$
$$w_{i2} = c, \qquad i = 1, \ldots, L-1,$$
$$w_{L1} = c + A,$$

where $c > 0, A > 0$. Thus, A is the additional cost incurred if the unit is allowed to become inoperative before being replaced; $\{W_t\}$ is the sequence of costs.

Either the discounted expected cost criterion $\Psi_R(i, \alpha)$ for some given α $(0 < \alpha < 1)$ or the expected average cost criterion ϕ_R may be of interest. The methods of Chapters 4 and 6 can be employed to find optimal replacement policies, depending on which criterion is selected.

However, in practice, it is frequently desirable to use simple replacement policies. For example, we speak of a *control limit* policy as one which always replaces the unit whenever the observed state is $i_0, i_0 + 1, \ldots, L$ and never replaces the unit in states $0, 1, \ldots, i_0 - 1$; state i_0 is the *control limit*. We shall show under certain conditions on $\{p_{ij}\}$ that there always exists a control limit policy that is optimal. We state

CONDITION A: The transition probabilities $\{p_{ij}\}$ are such that for every nondecreasing function $f(j), j = 0, \ldots, L$, the function

$$g(i) = \sum_{j=0}^{L} p_{ij} f(j), \qquad i = 0, \ldots, L-1$$

is also nondecreasing.

1 A Replacement Model

We also state

CONDITION B: The transition probabilities $\{p_{ij}\}$ are such that for each $k = 0, 1, \ldots, L$, the function

$$r_k(i) = \sum_{j=k}^{L} p_{ij}, \qquad i = 0, \ldots, L-1,$$

is nondecreasing.

Let us first show

LEMMA 1. Conditions A and B are equivalent.

Proof: Assume Condition A. Then, in particular, the function

$$f_k(j) = \begin{cases} 0, & j < k, \\ 1, & j \geq k, \end{cases}$$

is nondecreasing. But we have

$$g(i) = \sum_{j=0}^{L} p_{ij}\, f_k(j)$$

$$= \sum_{j=k}^{L} p_{ij}$$

$$= r_k(i),$$

and, hence, Condition B holds. Assume Condition B holds. Any nondecreasing function $f(j)$ can be expressed in the form

$$f(i) = \sum_{k=0}^{L} c_k\, f_k(i)$$

where $c_k \geq 0$, $k = 0, \ldots, L$ and $f_k(i)$ is defined above.
Then,

$$g(i) = \sum_{j=0}^{L} p_{ij}\, f(j)$$

$$= \sum_{j=0}^{L} p_{ij} \sum_{k=0}^{L} c_k\, f_k(j) \quad \text{(equation continued)}$$

$$= \sum_{k=0}^{L} c_k \sum_{j=0}^{L} p_{ij} f_k(j)$$

$$= \sum_{k=0}^{L} c_k \sum_{j=k}^{L} p_{ij}.$$

Since $c_k \geq 0$, and by hypothesis, $\sum_{j=k}^{L} p_{ij}$ is nondecreasing for each k, it follows that $g(i)$ is nondecreasing, proving the lemma.

The significance of Lemma 1 is that Condition A becomes a verifiable condition through the verification of condition B.

We now state:

THEOREM 1. *If Condition A (or B) holds, then there exists a control limit policy $R(\alpha)$ such that*

$$\Psi_{R(\alpha)}(i, \alpha) = \min_{R \in C} \Psi_R(i, \alpha), \qquad i = 0, \ldots, L.$$

Proof: Let $\Psi(i, \alpha, N) = \min_{R \in C} \sum_{t=0}^{N} \alpha^t E(W_t | Y_0 = i)$, $N = 0, \ldots, L$. Clearly, $\Psi(i, \alpha, 0)$ is a nondecreasing function of i. Assume $\Psi(i, 0, n)$ is nondecreasing in i for $0 \leq n \leq N$. Then since

$$\Psi(i, \alpha, N + 1)$$
$$= \min\left\{\alpha \sum_{j=0}^{L} p_{ij} \Psi(j, \alpha, N), \quad c + \alpha \sum_{j=0}^{L} p_{0j} \Psi(0, \alpha, N)\right\}, \qquad i \neq L$$
$$= c + A + \alpha \sum_{j=0}^{L} p_{0j} \Psi(0, \alpha, N), \qquad i = L,$$

from the induction hypothesis and Condition A, it follows that there exists an i^* such that

$$\Psi(i, \alpha, N + 1) = \alpha \sum_{j=0}^{L} p_{ij} \Psi(j, \alpha, N), \qquad i < i^*,$$
$$= c + \alpha \sum_{j=0}^{L} P_{0j} \Psi(j, \alpha, N), \qquad i^* \leq i < L,$$
$$= c + A + \alpha \sum_{j=0}^{L} p_{0j} \Psi(j, \alpha, N), \qquad i = L.$$

where $\Psi(i, \alpha, N + 1)$ is a nondecreasing function of i. Therefore, $\Psi(i, \alpha, N)$ is nondecreasing in i for $N = 0, 1, \ldots$. From Chapter 4, Theorem 1, we know that $\lim_{N \to \infty} \Psi(i, \alpha, N) = \min_{R \in C} \Psi_R(i, \alpha)$ is also nondecreasing in i. On repeating the argument using Condition A again, the theorem follows.

THEOREM 2. If Condition A (or B) holds, then there exists a control-limit policy R^* such that
$$\phi_{R^*}(i) = \min_{R \in C} \phi_R(i), \qquad i = 0, \ldots, L.$$

Proof: From Theorem 2, Chapter 3 and Corollary 1 to Theorem 1, Chapter 6, we need only consider policies in C_D. By Theorem 1 for each α ($0 < \alpha < 1$) there exists a control-limit policy $R(\alpha)$ that minimizes $\Psi_R(i, \alpha)$. Let $\{\alpha_v, v = 1, 2, \ldots\}$ be any sequence of discount factors such that $\lim_{v \to \infty} \alpha_v = 1$ and $R(\alpha_1) = R(\alpha_2) = \cdots = R^*$. Since there are at most a finite number of different control-limit policies, such a sequence exists. Let R be any policy in C_D. Since $R^* = R(\alpha_v)$ is optimal for α_v, we have
$$(1 - \alpha_v) \Psi_R(i, \alpha_v) \geq (1 - \alpha_v) \Psi_{R^*}(i, \alpha_v), \qquad v = 1, 2, \ldots.$$
Letting $v \to \infty$ and using Theorems 1, Appendix A, and 1(b), Appendix B, we obtain that
$$\begin{aligned}\Psi_R(i) &= \lim_{v \to \infty}(1 - \alpha_v)\Psi_R(i, \alpha) \\ &\geq \lim_{v \to \infty}(1 - \alpha_v)\Psi_{R^*}(i, \alpha) \\ &= \Psi_{R^*}(i), \qquad i = 0, \ldots, L.\end{aligned}$$
Therefore, R^* is optimal and the theorem is proved.

2 A Surveillance-Maintenance-Replacement Model

Consider a system, in use or in storage, which is deteriorating. Suppose that the deterioration occurs stochastically and that the condition of the system is known only if it is inspected, which is costly.

After inspection the manager of the system has two basic alternatives: (a) to replace the system or (b) to keep it. Under the second alternative he must decide the extent of repairs to be made and when to make the next inspection. If inspection is put off too long the system may fail in the interim, the consequence of which is an incurred cost which is a function of how long the system has been inoperative.

Let us suppose that the uninspected system evolves according to a Markov chain through the states $0, 1, \ldots, L$. The state 0, as before denotes a new system and L an inoperative system. Let $\{p_{ij}\}$ denote the matrix of transition probabilities with $p_{LL} = 1$ and $p_{iL} > 0$ for each i. Assume that when a replacement is made an instantaneous transition to state 0 takes place; when a repair is made an instantaneous transition takes place to one of the states, $1, \ldots, L - 1$ depending on the extent of the repairs. Replacements or repairs are only made at the time of inspections. Assume that $M < \infty$ denotes the upper bound on the number of periods that can elapse without an inspection.

Let c_i denote the cost of inspection when, in fact, the system is in state i. Let $r_{ij}, i = 1, \ldots, L, j = 0, \ldots, L - 1$ denote the cost to place the system in state j after observing the system to be in state i. In particular, r_{i0} is the cost to replace the system from state i. In addition we let $r_{L(m),j}, m = 1, \ldots, M$, denote the cost to place the system in state j from state L when prior to discovering the system in state L, the system has been in state L for m uninspected periods. This cost represents, in addition to the repair or replacement costs, the cost associated with undetected failure. For a criterion, we shall be interested in minimizing the expected average cost per unit time attributed to the surveillance-replacement-maintenance policy.

The Markovian decision process we shall work with is the process $\{Y_t, A_t, t = 0, 1, \ldots\}, Y_0 = 0$, where $\{Y_t, t = 0, 1, \ldots\}$ is the sequence of *observed* states and $\{A_t, t = 0, 1, \ldots\}$ is the sequence of actions taken. The state space I will consist of the states $0, 1, \ldots, L, L(1), \ldots, L(M)$, where $L(m), m = 1, \ldots, M$ are additional states with $L(m)$ denoting the fact that the system is observed to be in state L and has been in state L for m uninspected periods. At each state $i \in I$, an action $A_t = a_{jm}$ consists in placing the system in state $j, j = 0, 1, \ldots, L$ and deciding to

2 A Surveillance-Maintenance-Replacement Model

skip m ($m \leq M$) time periods before observing the system again. If the system is observed in one of the states $L, L(1), \ldots, L(M)$ we assume that $a_{j0}, j = 0, \ldots, L - 1$. are the only possible actions. We have as transition probabilities for the observed process

$$q_{ij}(a_{lm}) = p_{lj}^{(m+1)}$$

for each $i, j, l,$ and m.

Let

$$Z_{tijm} = 1, \quad \text{if } Y_t = i, \quad A_t = a_{jm},$$
$$= 0, \quad \text{otherwise},$$

and

$$\bar{Z}_{Tijm} = \frac{1}{T+1} \sum_{t=0}^{T} Z_{tijm}.$$

If J_T is the average cost up to time T (in real time) then, for each $i, j \in I$,

$$J_T = \frac{\sum_{t=0}^{t(T)} \sum_i \sum_j \sum_m (c_i + r_{ij}) Z_{tijm}}{\sum_{t=0}^{t(T)} \sum_i \sum_j \sum_m (1 + m) Z_{tijm} + \theta},$$

where $t(T)$ is the largest value of t such that the real time is less than or equal to T and θ in some positive integer less than or equal to M. We also have

$$J_T = \frac{\sum_i \sum_j \sum_m (c_i + r_{ij}) \bar{Z}_{t(T)ijm}}{\sum_i \sum_j \sum_m (1 + m) \bar{Z}_{t(T)ijm} + \theta(t(T))^{-1}}.$$

From the application to Theorem 5, Chapter 7 and since $t(T) \to \infty$ when $T \to \infty$, it is possible to select a policy $R \in C_S$ that minimizes $E \liminf_{T \to \infty} J_T$ (or $E \limsup_{T \to \infty} J_T$) over all $R \in C$. Since for any $R \in C_S$

$$\lim_{T \to \infty} \bar{Z}_{Tijm} = \lim_{T \to \infty} \frac{1}{T+1} \sum_{t=0}^{T} P\{Y_t = i, A_t = a_{jm}\}$$

with probability 1 (combine Theorems 4, Appendix A, and 5, Appendix B), the problem can, using the methods of Chapter 6, ultimately be put into the form: Choose $\{x_{ia_jm}\}$

to minimize

$$\frac{\sum_i \sum_j \sum_m (c_i + r_{ij}) x_{ia_jm}}{\sum_i \sum_j \sum_m (1 + m) x_{ia_jm}}$$

subject to

$$x_{ia_jm} \geq 0 \quad \text{for all} \quad i, j, m$$

and

$$\sum_j \sum_m x_{ia_jm} = \sum_k \sum_j \sum_m x_{ka_jm} q_{ki}(a_{jm}),$$

$$\sum_i \sum_j \sum_m x_{ia_jm} = 1.$$

The optimal policy is obtained by putting

$$D_{ia_jm} = \frac{x_{ia_jm}}{\sum_j \sum_m x_{ia_jm}}$$

for each i and a_{jm}.

The above problem involves minimizing a ratio of linear functions subject to linear constraints where the lower linear form is always positive. Any problem of this form can always be transformed to a linear programming problem. Namely, suppose we wish

to minimize

$$\frac{\sum_{i=1}^{n} c_i x_i}{\sum_{i=1}^{n} d_i x_i}$$

2 A Surveillance-Maintenance-Replacement Model

subject to

$$x_i \geq 0, \quad i = 1, \ldots, n,$$

$$\sum_{i=1}^{n} a_{ij} x_j = 0, \quad i = 1, \ldots, m,$$

$$\sum_{i=1}^{n} x = 1,$$

where $\sum_{i=1}^{n} d_i x_i > 0$ for all feasible $(x_1, \ldots, x_n)$. Set

$$z_i = \frac{x_i}{\sum_{i=1}^{n} d_i x_i}, \quad i = 1, \ldots, n,$$

$$z_{n+1} = \frac{1}{\sum_{i=1}^{n} d_i x_i}.$$

Then we can write the linear programming problem in $z_1, \ldots, z_{n+1}$

to minimize

$$\sum_{i=1}^{n} c_i z_i$$

subject to

$$z_i \geq 0, \quad i = 1, \ldots, n+1,$$

$$\sum_{j=1}^{n} a_{ij} z_j = 0, \quad i = 1, \ldots, m,$$

$$\sum_{i=1}^{n} z_i - z_{n+1} = 0,$$

$$\sum_{i=1}^{n} d_i z_i = 1.$$

Clearly a one-to-one correspondence exists between the feasible solutions of the two problems.

In the case of our problem, D_{ia_jm} can be obtained by solving the linear programming problem and setting

$$D_{ia_jm} = \frac{z_{ia_jm}}{\sum_j \sum_m z_{ia_jm}}$$

for each state i and decision a_{jm}. As argued in Chapters 4, 5, and 6, if the simplex method is used, it will turn out that for each state i, D_{ia_jm} will be equal to 1 for precisely one a_{jm} and equal to 0 for all others. Thus, a policy in C_D will be optimal among all policies.

3 AOQL of Continuous Sampling Plans

We have in mind a stream of items produced on a conveyor line. Each item, if inspected, could be classified as either a defective or a nondefective. Frequently, the stream is grouped into lots from which a sample is drawn and the lot accepted or rejected according to whether few or many defectives are found in the lot. Sampling in this manner in order to control the quality of output of the production system is called lot-by-lot sampling. However, often it is not feasible to group items in lots. When this is the case the continuous stream is sampled with defective items found replaced by nondefective ones. Sampling plans of this kind are called continuous sampling plans. We treat one of the simplest such plans—a version of what is usually referred to as Dodge plans. The plan proceeds as follows: Sample every item until m successive nondefective items are seen, at which time sample each item produced with probability f, $0 < f < 1$. When a defective is found, resort to 100% inspection. Then continue as before. Each defective item is replaced (from a pool of good items) by a nondefective item.

The out-going-quality (AOQ) for a given stream of items is defined as the least upper bound on the possible proportions of defective items that can pass through the inspection process. That is, for a given stream, because of the random inspections, the proportion of defective items passed (defined as the lim sup of the ratio of numbers of defective

3 AOQL of Continuous Sampling Plans

items passed to the number produced) is a random variable; the smallest value that exceeds this random variable with probability 1 is the AOQ.

A conservative measure of the effectiveness of a continuous sampling plan is the largest AOQ that can be obtained. If the AOQ can be kept within acceptable bounds no matter what the stream, then the plan is a satisfactory one. Accordingly, we define Average-Outgoing-Quality-Level as

$$AOQL = \sup_s AOQ,$$

where s denotes a production stream and the supremum is taken over all possible streams, or all possible streams in a given subset of streams.

The problem we consider here is that of computing the AOQL for a given m and f. For example we say the production process is in a state of "control" if each item has some fixed probability p of being defective. We can readily compute the AOQL under the restricted assumption that the process is in control. For let $\{Y_t\}$ be a stochastic process where Y_t denotes the state of the inspection system just before the tth item is inspected; that is, Y_t can be equal to $1, \ldots, m, m+1$, where $i, i = 1, \ldots, m$ denotes the state of being in 100% inspection with $i-1$ successive nondefectives already observed since being in 100% inspection, and $i = m+1$ denotes the state of being in partial inspection (sampling with probability f). Then $\{Y_t\}$ is a Markov chain with transition probabilities

$$p_{ii} = p, \qquad p_{i,i+1} = 1 - p, \qquad i = 1, \ldots, m,$$

$$p_{m+1,1} = pf, \qquad p_{m+1,m+1} = 1 - pf.$$

Using Theorem 4 of Appendix A the long range proportion π_{m+1} of time the inspection system is in state $m+1$ can be computed, from which we can obtain $AOQ = (1-f)p\pi_{m+1}$. In fact,

$$\pi_{m+1} = \frac{(1-p)^m}{f + (1-f)(1-p)^m}.$$

Hence,

$$AOQL = \max_{0 \le p \le 1} \frac{(1-f)p(1-p)}{f + (1-f)(1-p)^m}.$$

We use the Markovian decision process framework to compute the AOQL when the production process is not in a state of control. Let $\{Y_t, t = 1, 2, \ldots\}$ be as defined and consider the production process as the decision maker; that is, at the tth item, the production process, noting the state of the inspection system, decides to produce a nondefective or a defective. Thus, $A_t = 1$ or 2, 1 denoting a nondefective item and 2 a defective item. Let $\bar{Z}(m + 1, 2)$ denote the long range proportion of items for which $Y_t = m + 1$, $A_t = 2$ (or lim sup of long range proportion). Then using Theorem 5 of Appendix B (or the well-known law of large numbers), we can see that the long range proportion of passed defectives is $\bar{Z}(m + 1, 2)(1 - f)$ with probability 1. Thus, if $L_R = \inf \{L \mid P_R(\bar{Z}(m + 1, 2) \leq L) = 1\}$, the production process wants a policy R to maximize L_R; in so doing it obtains the AOQL, $L_R(1 - f)$. From the application to Theorem 5, Chapter 7 and because $\bar{Z}(m + 1, 2)$ can be maximized at an extreme point $\bar{H}$ (using the notation of Theorem 5 of Chapter 7), such a policy can be found in C_D. Thus, for certain states, $A_t = 1$ and for the others $A_t = 2$. Clearly, $A_t = 1$ if $Y_t = i, i = 1, \ldots, m$ and $A_t = 2$ if $Y_t = m + 1$. Under this policy $\{Y_t\}$ is a Markov chain with transition probabilities

$$p_{i,i+1} = 1, \qquad i = 1, \ldots, m,$$
$$p_{m+1,1} = f, \qquad p_{m+1,m+1} = 1 - f.$$

Using Theorem 4 of Appendix A one obtains that $\bar{Z}(m + 1, 2) = 1/(mf + 1)$ with probability 1 and hence, AOQL $= (1 - f)/(mf + 1)$.

4 A Sequential Search Problem

Suppose a hunted object moves about within a finite number of regions according to known probabilistic laws. A single searcher using a detection system whose effectiveness is a function of the amount of effort used and the region searched, checks one region at a time until the object is found, his effort budget is exhausted, or he decides it is "uneconomical" to continue. The problem is to find an optimal sequential search policy; that is, one which tells the searcher, at each point in

4 A Sequential Search Problem

time, whether to search, where to search, and how much effort to use in order to optimize a given effectiveness criterion.

More precisely, let us assume that there are regions labeled $1, \ldots, L$. At time t, $t = 0, 1, \ldots$, the object is in one of the regions. At that time the searcher selects one of the regions to search, say region i, and puts effort e into the search. If the searcher finds the object, the search is over; otherwise the object moves to region j at time $t+1$ with known probability h_{ij} ($i, j = 1, \ldots, L$). Initially, before the searcher has searched any regions, the object places itself in region j with probability h_{0j}. Thus, the movement of the object is a function of where the hunter searches.

We shall assume that the initial effort budget for the searcher is B, a nonnegative integer, and the efforts e expended are nonnegative integers. Also, in moving from region i to region j, the searcher uses up r_{ij} (a nonnegative integer) units of his effort budget.

We define the state space I of the Markovian decision process to be $I = \{i_b^k, i = 1, \ldots, L;\ k = 0, 1;\ b = 0, \ldots, B\} \cup \{O_B^0\}$. The state i_b^0 denotes that region i has just been searched, the object has not been found (indicated by the superscript 0), and b units of the budget remain for future use. The state i_b^1 denotes the fact that region i has just been searched, the object has been found (indicated by the superscript 1), and b units remain for future use. The state O_B^0 is a fictitious initial state from which a searcher begins his sequence of searches; O_B^0 can be interpreted as a base from which all search sequences begin. The class of states

$$T = \{i_0^0, \quad i = 1, \ldots, L\} \cup \{i_b^1, \quad i = 1, \ldots, L, \quad b = 0, \ldots, B\}$$

are states at which the search process terminates; in the first group, termination is due to lack of funds, in the second group, termination is due to the object having been found. When the process is in any of the states of T, the next point in time will find the process in O_B^0. Thus, the process returns to the initial state completing a cycle.

At each state of $I - T$ the searcher has a number of possible actions depending on the state. His possible actions at state i_b^0 are $\{a_{je}, j = 1, \ldots, L; e = 1, \ldots, b\}$, where a_{je} denotes the action of searching region

j and expending effort e. As stated in the previous paragraph, at each state of T the searcher has only one possible action—to return to $O_B{}^0$.

We let $v_j(e)$ denote the conditional probability of finding the object given that region j is searched, the object is in region j, and effort e is expended in the search. Thus, the laws of motion of the Markovian decision process can be written

$$q_{i_b{}^0 j_e^0{}_e}(a_{je}) = 1 - h_{ij} v_j(e),$$
$$q_{i_b{}^0 j_e^1{}_e}(a_{je}) = h_{ij} v_j(e), \qquad i = 0, \ldots, L; \quad 1 \leq e \leq b \neq 0,$$
$$q_{i_0{}^0 O_B{}^0} = 1,$$
$$q_{i_b{}^1 O_B{}^0} = 1, \qquad i = 0, \ldots, L, \quad b = 0, \ldots, B-1.$$

For costs, we have

$$w_{i_b{}^0 a_{je}} = e + r_{ij}, \qquad i, j = 0, \ldots, L; \quad e \leq b,$$
$$w_{i_0{}^0 O_B{}^0} = 0, \qquad i = 0, \ldots, L,$$
$$w_{i_b{}^1 O_B{}^0} = 0, \qquad i = 1, \ldots, L; \quad b = 1, \ldots, B-1.$$

Suppose we are interested in minimizing the expected cost to reach a state in T subject to the probability of a successful search being greater than or equal to a given number θ. The expected cost of reaching a state in T is the same as the expected cost of a return to state $O_B{}^0$. We can make use of Examples 2 and 3 of Chapter 7 as well as Theorem 2, Chapter 7, and Theorem 5 of Appendix A to formulate the problem as finding that policy $R \in C_S$

to minimize

$$\frac{1}{\pi_{O_B{}^0}(R)} \sum_{i \in I} \sum_{a} \pi_i(R) D_{ia}^R w_{ia}$$

subject to

$$\frac{1}{\pi_{O_B{}^0}(R)} \sum_{i \in T'} \pi_i(R) \geq \theta$$

where $T' = \{i_b{}^1, i = 1, \ldots, L; b = 0, \ldots, B-1\}$.

The fact that Theorem 2 of Chapter 7 applies is due to the fact that at most one ergodic class can exist in I for every $R \in C_D$.

Letting $x_{ia}^R = \pi_i(R) D_{ia}^R$, as in Chapter 5, the problem can be put into the form of finding $\{x_{ia}\}$

to minimize

$$\frac{\sum_{i \in I} \sum_a x_{ia} w_{ia}}{\sum_a x_{i_B{}^o a}}$$

subject to

$$x_{ia} \geqq 0, \quad a \in K_i, \quad i \in I,$$

$$\sum_a x_{ja} = \sum_i \sum_a x_{ia} w_{ia}, \quad j \in I,$$

$$\sum_i \sum_a x_{ia} = 1,$$

$$\sum_{i \in T'} \sum_a x_{ia} \geqq \theta \sum_a x_{O_B{}^o a}$$

The method of Section 2 of this chapter can then be used to transform this fractional linear programming problem to an ordinary linear programming problem and then obtain from its solution the optimal policy in C_S.

Other criteria could be formulated with the same methods for obtaining the optimal policy prevailing. For example we could simply maximize the probability of a successful search.

5 A Stochastic Traveling Salesman Problem

In the well-known traveling salesman problem there are $L + 1$ cities $0, \ldots, L$. There is a cost of w_{ij} to travel from city i to city j. The problem consists of finding a route, starting and ending at city 0 through all cities, which has minimum total cost. Clearly, a minimal route goes through each city, not counting the initial time in city 0, exactly once.

In the stochastic traveling salesman problem we allow the routes to be determined by chance with the constraint that in a route starting from 0 and returning to 0 the expected number of visits to city i be exactly one. We seek a random mechanism for generating the routes to have the property that the expected total cost of a route be minimized.

To approach this problem with the tools of Markovian decision processes, let I denote the states $0, \ldots, L$. At each state suppose there are actions $a = 0, \ldots, L$; that is, action $a = i$ is an effort to move the process to state i. However, for technical reasons, we assume that if action $a = j$ is taken in state i, then for $i, j \in I$,

$$q_{ij}(a = j) = 1 - \varepsilon, \qquad q_{ij'}(a = j) = \frac{\varepsilon}{L}, \qquad j' \neq j,$$

for some small positive number ε. We assume $\varepsilon > 0$ in order that I be irreducible for every policy $R \in C_D$. The cost of being in state i and taking action $a = j$ is w_{ij}.

Using the results of Example 2 of Chapter 7 and Theorem 2 of Chapter 7 and Theorem 5 of Appendix A we can formulate the stochastic traveling problem as finding that policy $R \in C_S$ such that

$$\frac{1}{\pi_0(R)} \sum_i \sum_j \pi_i(R) D_{ij}^R w_{ij}$$

is minimized subject to

$$\frac{\pi_i(R)}{\pi_0(R)} = 1, \qquad i \in I.$$

From the latter constraints we obtain that $\pi_0(R) = 1/(L + 1)$. Thus, using the transformation $x_{ij} = \pi_i(R) D_{ij}^R$ of Chapter 5, we can formulate the problem as that of finding $\{x_{ij}\}$

to minimize

$$\sum_i \sum_j x_{ij} w_{ij}$$

Bibliographical Remarks

subject to

$$x_{ij} \geqq 0, \quad i, j \in I,$$

$$\sum_j x_{ij} = \sum_k \sum_a x_{ka} q_{kj}(a), \quad i \in I,$$

$$\sum_i \sum_j x_{ij} = 1,$$

$$\sum_j x_{ij} = \frac{1}{L+1}, \quad i \in I.$$

Since, for each $j \in I$,

$$\sum_k \sum_a x_{ka} q_{kj}(a) = \sum_k x_{kj}(1-\varepsilon) + \frac{1}{L+1} \sum_k \sum_{a \neq j} x_{ka} \varepsilon$$

$$= \sum_k x_{kj}(1-\varepsilon) + \frac{1}{L+1} \sum_k \left(\frac{1}{L+1} - x_{kj}\right)\varepsilon$$

$$= \sum_k x_{kj}\left(1 - \frac{\varepsilon}{L+1}\right) + \frac{\varepsilon}{L+1},$$

we can rewrite the equations as

$$\sum_i x_{ij} = \frac{1-\varepsilon}{L+1-\varepsilon}, \quad j \in I,$$

$$\sum_j x_{ij} = \frac{1}{L+1}, \quad i \in I.$$

As in Chapter 5, we then let $D_{ij}^R = x_{ij}/\sum_j x_{ij}$ be the optimal policy, where $\{x_{ij}\}$ is the optimal solution to the linear programming problem.

Bibliographical Remarks

The material of Section 1 appears in Derman [20]. Further results are obtained by Kolesar [41]. A similar problem maximizing the expected time between replacements subject to a probability bound on failure is treated by Derman [21] and is extended by Kolesar [40] to show the control limit optimality for this case.

Section 2 is found in Klein [38]. We make use of our results in Chapter 7 to reduce the problem to policies in C_S. In the original paper, only policies in C_S were permitted. The transformation to the linear programming problem appears in Derman [15]. See Charnes and Cooper [9] for more general treatment of "linear fractional programming" problems.

The calculation of AOQL's for continuous sampling plans via Markovian decision processes has been treated more generally by White [54]. The out-of-control AOQL for the Dodge continuous sampling plans was first computed by Lieberman [42]. For Dodge's original paper see [27].

The application of Markovian decision models to the sequential search problem is due to Klein [39].

We refer the reader to Derman and Klein [23] for the original version of the stochastic traveling salesman problem.

Appendix A. Markov Chains

A sequence of random variables $\{Y_t = t = 0, 1, \ldots\}$ whose range is a finite state space I is called a *finite-state Markov chain* with *stationary transition probabilities* $\{p_{ij}, i, j \in I\}$ if

$$P\{Y_{t+1} = j \mid Y_0 = y_0, \ldots, Y_{t-1} = y_{t-1}, Y_t = i\} = p_{ij}$$

for every $y_0, \ldots, y_{t-1}, i, j \in I$, and $t = 0, 1, \ldots$. The parameters $\{p_{ij}\}$ are called *one-step transition probabilities*. More generally, it follows that

$$P\{Y_{s+t} = j \mid Y_0 = y_0, \ldots, Y_{s-1} = y_{s-1}, Y_s = i\} = P\{Y_{s+t} = j \mid Y_s = i\}$$
$$= P\{Y_t = j \mid Y_0 = i\}$$
$$= p_{ij}^{(t)}$$

for every $y_0, \ldots, y_{s-1} > i, j \in I, s = 0, 1, \ldots, t = 1, 2, \ldots$. We refer to $\{p_{ij}^{(t)}\}$ as the *t-step transition* probabilities. For $t = 0$, we define $p_{ij}^{(0)} = \delta_{ij}$, the Kronecker delta. If $t = 1$, $p_{ij} = p_{ij}^{(1)}$.

The fundamental relationships connecting the transition probabilities are the Chapman–Kolmogorov equations

$$p_{ij}^{(t+s)} = \sum_{k \in I} p_{ik}^{(t)} p_{kj}^{(s)}, \qquad i, j \in I, \tag{A.1}$$

holding for all $s, t \geqq 0$. In particular one gets the recursive formulas

$$\begin{aligned} p_{ij}^{(t+1)} &= \sum_{k \in I} p_{ik}^{(t)} p_{kj} \\ &= \sum_{k \in I} p_{ik} p_{kj}^{(t)}. \end{aligned} \tag{A.2}$$

As a consequence of Eqs. (1) and (2) certain fundamental results can be obtained. An important result is:

THEOREM 1. *For all* $i, j \in I$,

$$\pi_{ij} = \lim_{T \to \infty} \frac{1}{T+1} \sum_{t=0}^{T} p_{ij}^{(t)}$$

exists and satisfies

$$\pi_{ij} \geqq 0,$$

$$\pi_{ij} = \sum_{k \in I} \pi_{ik} p_{kj} = \sum_{k \in I} p_{ik} \pi_{kj} = \sum_{k \in I} \pi_{ik} \pi_{kj}, \tag{A.3}$$

$$\sum_{j \in I} \pi_{ij} = 1.$$

It is convenient to classify the various states of I. If there exist positive integers t_1 and t_2 such that $p_{ij}^{(t)} > 0$ and $p_{ji}^{(t)} > 0$, we say that i and j *communicate*. If I' is a set of states all of which communicate and no set containing I' possesses states all of which communicate, then I' is called a class (or *ergodic* class). In general I may be divided into a number of classes and some states which do not belong to any classes. When I consists of one class $I' = I$, then we say I is *irreducible*.

States that belong to some class are called *recurrent*. States that do not belong to any class are called *transient*. Transient states can be characterized in a different way. A state i is transient if there exists a state j and an integer t such that $p_{ij}^{(t)} > 0$; but $p_{ji}^{(t)} = 0$, $t = 0, 1, \ldots$. If j is transient then $\lim_{t \to \infty} p_{ij}^{(t)} = \pi_{ij} = 0$ for all $i \in I$.

Let I' denote a set of transient states and denote by Q the matrix $\{p_{ij}, i, j \in I'\}$:

THEOREM 2. The matrix $I - Q$ has an inverse, namely,
$$(I - Q)^{-1} = I + Q + Q^2 + \cdots$$
$$< \infty.$$

A result similar to Theorem 2 is stated in:

THEOREM 3. If Q is any matrix of positive elements whose row sums are less than or equal to unity and ϵ is any number $0 \leq \alpha < 1$, then
$$(I - \alpha Q)^{-1} = I + \alpha Q + \alpha^2 Q^2 + \cdots$$
$$< \infty.$$

The following theorem enables one to evaluate π_{ij} when j is recurrent.

THEOREM 4. Let I' and I'' be classes of recurrent states. If $i \in I'' \neq I'$, then $\pi_{ij} = 0$, $j \in I'$. If $i \in I'$, $j \in I'$, then $\pi_{ij} = \pi_j$, independent of i, where $\{\pi_j, j \in I'\}$ uniquely satisfy
$$\pi_j > 0,$$
$$\pi_j = \sum_{i \in I'} \pi_i p_{ij},$$
$$\sum_{j \in I'} \pi_j = 1.$$

If i is transient and $j \in I'$, then
$$\pi_{ij} = P\{Y_t \in I' \text{ for some } t > 0 \mid Y_0 = i\} \pi_j.$$

Let $Z_t = 1$ if $Y_t = j$, or zero otherwise. If $i \in I'$, $j \in I'$, and $Y_0 = i$, then

$$\lim_{T \to \infty} \frac{1}{T+1} \sum_{t=0}^{T} Z_t = \pi_j$$

with probability 1.

Further useful results can be summarized in:

THEOREM 5. Suppose I' is a class of recurrent states, suppose $j \in I'$ and $\tau = \min\{t \geq 1 \mid Y_t = j\}$. Then

$$E\{\tau \mid Y_0 = j\} = \frac{1}{\pi_j}.$$

Also, if $\{w_i, i \in I\}$ is a set of numerical values and $u_j = E\left\{\sum_{t=0}^{\tau-1} W_t \mid Y_0 = j\right\}$, where $W_t = w_i$ if $Y_t = i$, then

$$u_j = \frac{1}{\pi_j} \sum_{i \in I'} \pi_i w_i.$$

Let $f_{ij}^{(t)} = P\{Y_t = j, Y_n \neq j \text{ for some } 1 \leq n < t \mid Y_0 = i\}$. Then $\sum_{t=1}^{\infty} f_{ij}^{(t)} = P\{Y_t = j \text{ for some } t \geq 1 \mid Y_0 = i\}$. We let $m_{ij} = \sum_{t=1}^{\infty} t f_{ij}^{(t)}$, which can be interpreted as the mean first-passage time from state i to state j when $\sum_{t=1}^{\infty} f_{ij}^{(t)} = 1$.

The following theorem is useful:

THEOREM 6. If $P\{Y_t = j \text{ for some } t \geq 1 \mid Y_0 = i\} > 0$ for all $i \in I$, then $\sum_{t=1}^{\infty} f_{ij}^{(t)} = 1$ for all $i \in I$, and furthermore $m_{ij} < \infty$ for all $i \in I$.

Bibliographical Notes

All results stated here can be found with proofs in Chung [11], Feller [31], and/or Kemeny and Snell [37].

Appendix B. Some Theorems from Analysis and Probability Theory

Let $\{a_n, n = 0, 1, \ldots\}$ be a sequence of real numbers and let

$$f(x) = \sum_{n=0}^{\infty} a_n x^n, \qquad 0 \leqq x \leqq 1.$$

THEOREM 1. (a) If

$$\sum_{n=0}^{\infty} a_n = A < \infty,$$

then
$$\lim_{x \to 1} f(x) = A.$$

(b) If
$$\lim_{N \to \infty} \frac{1}{N} \sum_{n=0}^{N} a_n = A < \infty,$$

then
$$\lim_{x \to 1}(1 - x)f(x) = A.$$

(c) Let
$$\lim_{N \to \infty} \sup \frac{1}{N} \sum_{n=0}^{N} a_n = A,$$

then
$$\lim_{x \to 1} \sup (1 - x)f(x) \leq A.$$

Let S be any space of points. If there is a function $\rho(x, y)$ of x and y in S, satisfying

$\rho(x, x) = 0,$ $\qquad x \in S,$

$\rho(x, y) = \rho(y, x),$ $\qquad x, y \in S,$

$\rho(x, z) \leq \rho(x, y) + \rho(x, z),$ $\qquad x, y, z \in S,$

then ρ is called a *metric* and S is called a *metric space* (with metric ρ). We say a set of points $\{x_n\} \in S$ converges to a point $x \in S$ if $\lim_{n \to \infty} \rho(x_n, x) = 0$.

A space S is said to be compact if for every sequence of points $\{x_n\} \in S$ there exists a subsequence $\{x_{n_v}, v = 1, \ldots\}$ and a point $x \in S$ such that $\{x_{n_v}, v = 1, \ldots\}$ converges to x.

Let $f(x)$ be any function on S. A well-known fact about continuous functions over compact spaces is:

Analysis, Probability Theory Theorems

THEOREM 2. If f is a continuous function on S and S is compact, then there exists a point $x^* \in S$ such that $\min_{x \in S} f(x) = f(x^*)$.

Let $S_1, S_2, \ldots$ be a denumerable collection of spaces. By the product space $\prod S_i$, we mean the set of all possible sequences $x = \{x_i, i \in S_i, i = 1, 2, \ldots\}$. We shall say that a sequence $\{x^{(n)}, n = 1, 2, \ldots\}$ of points in $\prod S_i$ converges to a point $x \in \prod S_i$ if $\lim_{n \to \infty} \rho_i(x_i^{(n)}, x_i) = 0$, for every $i = 1, 2, \ldots$ where $x_i^{(n)}$ is the ith component of $x^{(n)}$, x_i is the ith component of x and ρ_i is the metric of S_i.

In Chapter 3 we used:

THEOREM 3. If S_i is compact for each $i = 1, 2, \ldots$, then $\prod S_i$ is also compact.

Proof: Let $\{x^{(n)}\}$, be any sequence of points in $\prod S_i$. Since S_1 is compact there exists a subsequence $\{n_v^1, v = 1, \ldots\}$ and a point $x_1 \in S_1$ such that $\lim_{v \to \infty} \rho_1(x^{n_v^1}, x_1) = 0$. Since S_2 is compact, there exists a subsequence $\{n_v^2, v = 1, \ldots\}$ of $\{n_v^1\}$ and a point $x_2 \in S_2$ such that $\lim_{v \to \infty} \rho_2(x_2^{n_v^2}, x_2) = 0$. Continuing, for each i there exists a subsequence $\{n_v^i, v = 1, \ldots\}$ of $\{n_v^{i-1}, v = 1, \ldots\}$ and a point $x_i \in S_i$ such that $\lim_{v \to \infty} \rho_i(x_i^{n_v^i}, x_i) = 0$. Now let $n_v = n_v^v, v = 1, \ldots,$ and $x = (x_1, x_2, \ldots)$ $\in \prod S_i$. Then, since $\{n_v, v = 1, \ldots\}$ is a subsequence of $\{n_v^i\}$ for each i, we have $\lim_{v \to \infty} \rho_i(x_i^{n_v}, x_i) = 0$. Thus, $\prod S_i$ is compact.

Let $\{Y_t, t = 1, 2, \ldots\}$ be any sequence of random variables. We use the following:

THEOREM 4. (a) If $Y_t \geq 0, t = 1, 2, \ldots$, then

$$E \sum_{t=1}^{\infty} Y_t = \sum_{t=1}^{\infty} E Y_t.$$

(b) If $E \sum_{t=1}^{\infty} |Y_t| < \infty$ or $\sum_{t=1}^{\infty} E|Y_t| < \infty$, then

$$E \sum_{t=1}^{\infty} Y_t = \sum_{t=1}^{\infty} EY_t.$$

THEOREM 5. If

$$\sum_{t=1}^{\infty} \frac{\operatorname{Var} Y_t}{t^2} < \infty,$$

then

$$\lim_{T \to \infty} \frac{1}{T} \sum_{t=1}^{T} [Y_t - E(Y_t | Y_1, \ldots, Y_{t-1})] = 0,$$

with probability 1.

Suppose $\{Y_t, t = 1, 2, \ldots\}$ is a sequence of random variables with associated random times $\tau_1, \tau_2, \ldots$ such that the sets of random variables

$$\{Y_1, \ldots, Y_{\tau_1}\}, \quad \{Y_{\tau_1+1}, \ldots, Y_{\tau_2}\}, \ldots$$

are independent and have the same probability laws. It is assumed (letting $\tau_0 = 0$) that $\{\tau_v - \tau_{v-1}, v = 1, \ldots\}$ are independent and identically distributed with $\tau_v - \tau_{v-1}$ being determined by conditions on $Y_{\tau_{v-1}+1}, \ldots, Y_{\tau_v}$. Such a sequence we call a *recurrent* event process; that is, some event occurs at times $\tau_1, \tau_2, \ldots$ which has the effect of starting the process $\{Y_t, t = 1, 2, \ldots\}$ anew. For example, if $\{Y_t, t = 0, 1, \ldots\}$ is a Markov chain with $Y_0 = i$, a recurrent state, $\tau_1, \tau_2, \ldots$ may be the successive times at which $Y_t = i$.

Let $\{C_v, v = 1, 2, \ldots\}$ be a sequence defined by

$$C_v = C(Y_{\tau_{v-1}+1}, \ldots, Y_{\tau_v}), \quad v = 1, 2, \ldots;$$

that is, C is a function of the random variables in the vth "cycle" in the recurrent event process. The sequence $\{C_v, v = 1, 2, \ldots\}$ consists of independent and identically distributed random variables.

One can show:

THEOREM 6. If $EC_v < \infty$ and $E(\tau_v - \tau_{v-1}) < \infty$ and if $v(t)$ is the largest v such that $\tau_v \leq t$, then

$$\lim_{t \to \infty} E \frac{\sum_{v=1}^{\tau_{v(t)}} C_v}{t} = \lim_{t \to \infty} E \frac{\sum_{v=1}^{\tau_{v(t)}+1} C^v}{t}$$

$$= \frac{EC_v}{E(\tau_v - \tau_{v-1})}.$$

If we interpret C_v to be a cost associated with the vth cycle defined by the function C, then Theorem 6 asserts that the expected average cost per unit time is equal to the ratio of the expected cost of a cycle to the expected length of a cycle. See Theorem 5 of Appendix A for application in the Markov chain context. We can make the same assertion for the average cost per unit time, the limit existing and taking on the same value with probability one.

Bibliographical Notes

Theorem 1 summarizes several well-known Abelian theorems. See, for example, Widder [55].

Theorem 2 can be found in any text on real analysis. Theorem 3 is a special case of Tychonov's theorem. We only need $\prod S_i$ to be the product space of a denumerable number of spaces S_i. Tychonov's theorem holds for any product space of compact spaces. The theorem asserts that $\prod S_i$ is compact in the product topology, equivalent to the topology we implicitly defined.

Theorem 4(a) is a consequence of the Lebesgue monotone convergence theorem, and (b) is a consequence of the theorem of Fubini.

Theorem 5 is a strong law of large numbers for dependent random variables. (See Loeve [44], p. 387.)

See Feller [31] for a treatment of recurrent event processes.

Appendix C. Convex Sets and Linear Programming

Let E denote an n-dimensional Euclidean space. A set $S \subset E$ is said to be convex if whenever $x \in S$, $y \in S$, then $\alpha x + (1 - \alpha)y \in S$ for every real number α, $0 \leq \alpha \leq 1$.

A fact we use about convex sets is:

THEOREM 1. If S is a closed convex set and $x = (x_1, \ldots, x_n) \notin S$ then there exists a set of real numbers $\{w_1, \ldots, w_n\}$ such that for every

$$y = (y_1, \ldots, y_n) \in S, \quad \sum_{i=1}^{n} w_i y_i > \sum_{i=1}^{n} w_i x_i.$$

The *convex hull* of a set $T \subset E$ is the smallest convex set S such that $S \supset T$. The *closed convex hull* of T is the smallest closed convex set containing T; namely, S, the closure of S.

An extreme point of a convex set S is any point $x \in S$ such that there does not exist points $y \in S$, $z \in S$ distinct from x for which $x = \alpha y + (1 - \alpha)z$ for some α, $0 < \alpha < 1$.

THEOREM 2. Let P denote the set of extreme points of a closed convex set S. Then every point $x \in S$ can be expressed in the form $x = \sum_{i=1}^{k} \alpha_i z_i$, where $z_i \in P$, $i = 1, \ldots, k$, $0 \leq \alpha_i \leq 1$, and $\sum_{i=1}^{k} \alpha_i = 1$.

A function $f(\cdot)$ over a convex set S is concave (convex) if for every $x_1 \in S$, $x_2 \in S$, and α, $0 \leq \alpha \leq 1$,

$$f(\alpha x_1 + (1 - \alpha)x_2) \geq (\leq) \alpha f(x_1) + (1 - \alpha)f(x_2).$$

THEOREM 2a. If $f(\cdot)$ is concave (convex) over a closed convex set S and if it achieves its minimum (maximum), then $f(\cdot)$ is minimized (maximized) at an extreme point of S.

A linear programming problem is a mathematical optimization problem that can be formulated in one of several standard forms. A problem given in one form can always be translated into any one of the other standard forms. The following is one form: To find variables $x_1, \ldots, x_n$

to minimize

$$\sum_{j=1}^{n} c_j x_j$$

subject to the constraints

$$\sum_{j=1}^{n} a_{ij} x_j = b_i, \quad i = 1, \ldots, m \quad \text{(C.1)}$$

and

$$x_j \geq 0, \quad j = 1, \ldots, n, \quad \text{(C.2)}$$

Convex Sets and Linear Programming

where $\{c_j, j = 1, \ldots, n\}$, $\{a_{ij}, i = 1, \ldots, m;\ j = 1, \ldots, n\}$, and $\{b_i, i = 1, \ldots, m\}$ are given real valued constants.

A set of values that satisfy (1) and (2) is called a feasible solution to the linear programming problem. The set of all feasible solutions is a closed convex set possessing a finite number of extreme points. The linear expression $\sum_{j=1} c_j x_j$ is called the objective function. A feasible solution that minimizes the objective function is called an optimal solution. Depending on circumstances, there may exist one, many, or no optimal solutions. However, if at least one optimal solution exists then there exists an extreme point in the convex set of feasible solutions which is an optimal solution. We can assert:

THEOREM 3. *An extreme point of the set of feasible solutions of the linear programming problem stated in the given form has at most m positive components.*

Since there is an extreme point which is optimal whenever an optimal solution exists, computational methods may restrict themselves to searching the extreme point for an optimal solution. The *simplex method* is a general method for solving linear programming problems and which yields an extreme point optimal solution. A linear programming problem must be translated to the above form in order to use the simplex method.

A problem related to the one stated above is: To find variables $v_1, \ldots, v_m$

to maximize
$$\sum_{i=1}^{m} b_i v_i$$

subject to the constraints

$$\sum_{i=1}^{m} a_{ij} \leq c_j, \quad j = 1, \ldots, n,$$

where

$$\{c_j, \ j = 1, \ldots, n\}, \ \{b_i, \ i = 1, \ldots, m\},$$
$$\{a_{ij}, \ i = 1, \ldots, m; \ j = 1, \ldots, n\}$$

are as before. This second problem is a linear programming problem in a different form. This problem is called the *dual* problem to the first problem; the first, is referred to as the *primal* problem. Every linear programming problem, regardless of the form in which it is stated, has a well-defined dual problem, its form depending on the form of the primal. In particular, the dual of the dual problem is always the primal problem; thus, there is no difference which problem is originally referred to as the primal.

The central relationship between primal and dual problems can be stated as:

THEOREM 4. Suppose an optimal solution $x_1^*, \ldots, x_n^*$ exists to the primal problem, then an optimal solution $v_1^*, \ldots, v_m^*$ exists to the dual problem and

$$\sum_{j=1}^{n} c_j x_j^* = \sum_{i=1}^{m} b_i v_i^*.$$

We also make use of the following:

THEOREM 5. Let $u_j = c_j - \sum_{i=1}^{m} a_{ij} v_i, j = 1, \ldots, n$. A necessary and sufficient condition that feasible solutions $x_1, \ldots, x_n$ and $v_1, \ldots, v_m$ both be optimal for their respective problems is that

$$\sum_{j=1}^{n} x_j u_j = 0.$$

Bibliographical Notes

Our sources for information on convex sets and linear programming include Karlin [36] and Dantzig [12].

References

1. Balinski, M. L., On Solving Discrete Stochastic Decision Problems. U.S. Navy Supply System Research Study 2. Mathematica, Princeton, New Jersey, 1961.
2. Bellman, R., A Markovian decision process, *J. Math. Mech.* **6**, 679–684 (1957).
3. Bellman, R., "Dynamic Programming." Princeton Univ. Press, Princeton, New Jersey, 1957.
4. Bellman, R. and Lasalle, J. P., "On Non-Zero Sum Games and Stochastic Processes." Rand McNally, Chicago, Illinois, 1949.
5. Bellman, R. and Blackwell, D., "On a Particular Non-Zero Sum Game." Rand McNally, Chicago, Illinois, September, 1949.
6. Blackwell, D., Discrete dynamic programming, *Ann. Math. Statist.* **33**, 719–726 (1962).

7. Blackwell, D., Discounted dynamic programming, *Ann. Math. Statist.* **36**, 226–235 (1965).
8. Breiman, L., Stopping rule problems, *in* "Applied Combinatorial Mathematics," Chapter 10. Wiley, New York, 1964.
9. Charnes, A. and Copper, W. W., Programming with fractional functionals: I, Linear fractional programming *Naval Res. Logist. Quart.* **9**, Nos. 3 & 4, 181–186 (1962).
10. Chow, Y. S. and Robbins, H., A martingale system theorem and applications, *in* Proc. Berkeley Symp., Math. Statist. Prob. 4th, pp. 93–104. Univ. of California Press, Berkeley, California, 1961.
11. Chung, Kai Lai, "Markov Chains with Stationary Transition Probabilities." Springer, Berlin, 1960.
12. Dantzig, G., "Linear Programming and Extensions." Princeton Univ. Press, Princeton, New Jersey, 1963.
13. Denardo, E. V. and Fox, B. L., Multichain Markov renewal programs, *Siam J. Appl. Math.* **16**, 468–487 (1968).
14. D'Epenoux, F., Sur un probleme de production et de stockage dans l'a léatoire, *Rev. Francaise Informat. Recherche Opérationnelle* **14**, 3–16 (1960). [English transl.: *Mgt. Sci.* **10**, 98–108 (1963).]
15. Derman, C., On sequential decisions and Markov chains, *Mgt. Sci.* **9**, 16–24 (1962).
16. Derman, C., Stable sequential control rules and Markov chains, *J. Math. Anal. Appl.* **6**, 257–265 (1963).
17. Derman, C., Markovian sequential control processes—denumerable state space, *J. Math. Anal. Appl.* **10**, 295–302 (1965).
18. Derman, C., On sequential control processes, *Ann. Math. Statist.* **35**, 341–349 (1964).
19. Derman, C., Denumerable state Markovian decision processes—average cost criterion, *Ann. Math. Statist.* **37**, 1545–1554 (1966).
20. Derman, C., On optimal replacement rules when changes of state are Markovian, *in* "Mathematical Optimization Techniques" (R. Bellman, ed.), Chapter 9, pp. 201–210. Univ. of California Press, Berkeley, California, 1963.
21. Derman, C., Optimal replacement under Markovian deterioration with probability bounds on failure, *Mgt. Sci.* **9**, 478–481 (1963).
22. Derman, C. and Klein, M., Some remarks on finite horizon Markovian decision models, *Operations Res.* **13**, 272–278 (1965).
23. Derman, C. and Klein, M., Surveillance of multi-component systems: A stochastic travelling salesman problem, *Naval Res. Logist. Quart.* **13**, 103–111 (1966).
24. Derman, C. and Sacks, J., Replacement of periodically inspected equipment (An optimal optimal stopping rule), *Naval Res. Logist. Quart.* **7**, 597–607 (1960).
25. Derman, C. and Strauch, R., A note on memoryless rules for controlling sequential control processes, *Ann. Math. Statist.* **37**, 276–278 (1966).
26. Derman, C. and Veinott, Jr., A. F., A solution to a countable system of equations

References

arising in Markovian decision processes, *Ann. Math. Statist.* **38**, 582–584 (1967).

27. Dodge, H. F., A sampling inspection plan for continuous production, *Ann. Math. Statist.* **14**, 264–279 (1943).
28. Doob, J. L., "Stochastic Processes." Wiley, New York, 1953.
29. Dynkin, E. B., The optimum choice of instant for stopping a Markov process, *Soviet Math. Dokl.* [*English Transl.*] **4**, 627–629 (1963).
30. Eaton, J. H. and Zadeh, L. A., Optimal pursuit strategies in discrete state probabilistic systems, *Trans. ASME Ser. D., J. Basic Engineering* **84**, 23–29 (1962).
31. Feller, W., "An Introduction to Probability Theory and Its Applications," 3rd ed. Wiley, New York, 1968.
32. Fisher, L. and Ross, S., An example in denumerable decision processes, *Ann. Math. Statist.* **39**, 674–676 (1968).
33. Gillette, D., Stochastic games with zero stop probabilities, *Ann. Math. Studies* **3**, 179–186 (1957).
34. Howard, R. A., "Dynamic Programming and Markov Processes." Technology Press, Cambridge, Massachusetts, and Wiley, New York, 1960.
35. Karlin, S., Structure of dynamic programming, *Naval Res. Logist. Quart.* **2**, 285–294 (1955).
36. Karlin, S., "Mathematical Methods and Theory in Games Programming and Economics," Vol. 1. Addison–Wesley, Reading, Massachusetts, 1959.
37. Kemeny, J. G. and Snell, J. L., "Finite Markov Chains." Van Nostrand, Princeton, New Jersey, 1960.
38. Klein, M., Inspection-maintenance-replacement schedules under Markovian deterioration, *Mgt. Sci.* **9**, 25–32 (1962).
39. Klein, M., Note on sequential search, *Naval Res. Logist. Quart.* **15**, 469–475 (1968).
40. Kolesar, P., Randomized replacement rules which maximize the expected cycle length of equipment subject to Markovian deterioration, *Mgt. Sci.* 867–876 (1967).
41. Kolesar, P., Minimum cost replacement under Markovian deterioration, *Mgt. Sci.* **12**, 694–766 (1966).
42. Lieberman, G. J., A note on Dodge's continuous inspection plans, *Ann. Math. Statist.* **24**, 480–484 (1953).
43. Liggett, T. M. and Lippman, S. A., Stochastic Games with Perfect Information. Working Paper 142, Western Management Science Institute, October (1968).
44. Loeve, M., "Probability Theory." Van Nostrand, Princeton, New Jersey, 1960.
45. Maitra, A., Dynamic programming for countable state systems, *Sankhya Ser. A.* **27**, Parts 2, 3, & 4, 241–248 (1965).
46. Manne, A. S., Linear programming and sequential decisions, *Mgt. Sci.* **6**, 259–267 (1960).
47. Miller, B. L. and Veinott, A. F., Discrete dynamic programming with small interest rate, *Ann. Math. Statist.* **40**, 366–370 (1969).
48. Shapley, L. S., Stochastic games, *Proc. Nat. Acad. Sci. U.S.* **39** (1953).

49. Snell, J. L., Applications of martingale system theorems, *Trans. Amer. Math. Soc.* **73**, 293–312 (1952).
50. Strauch, R. and Veinott, A. F., "A Property of Sequential Control Processes." Rand McNally, Chicago, Illinois, 1966.
51. Taylor, H., Optimal stopping in a Markov process, *Ann. Math. Statist.* **39**, 1333–1344 (1968).
52. Taylor, H., Optimal Stopping of Averaged Brownian Motion. Tech. Rept., Dept. Operations Res. Cornell Univ., Ithaca, New York, November (1967).
53. Veinott, A. F., Jr., Discrete dynamic programming with sensitive discount optimality criteria, *Ann. Math. Statist.*, **40**, 1635–1660 (1969).
54. White, L. S., Markovian decision models for the evaluation of a large class of continuous inspection plans, *Ann. Math. Statist.* **36**, 1408–1420 (1965).
55. Widder, D., "Laplace Tranform." Princeton Univ. Press, Princeton, New Jersey, 1946.

Index

A

Abelian theorems, 147
Actions, number of, set of, sequence of, 3
Average cost
 criterion, 6
 problem, 6, 20, 25

B

Backward induction, 11
Balinski, M. L., 84, 153
Bellman, R., 7, 8, 17, 50, 84, 153
Blackwell, D., 7, 8, 17, 33, 50, 83, 153, 154
Breiman, L., 116, 117, 119, 154

C

Chapman–Kolmogorov equations, 140
Charnes, A., 138, 154
Chow, Y. S., 116, 117, 154
Chung, K. L., 142, 154
Communicating states, 140
Compact metric spaces, 144
Compactness of policy space, 20
Concave functions, 150
Continuous sampling plans, 130
 AOQ of, 130
 AOQL of, 131
 Dodge type, 130
Control limit, 122
Control limit policy, 122

158   Index

Convergence of policies, 20
Convex functions, 150
Convex hull, closed convex hull, 150
Convex sets, 149
   extreme points of, 150
Cooper, W. W., 138, 154
Cost structure, 4

D

Dantzig, G., 152, 154
Denardo, E. V., 84, 154
D'Epenoux, F., 8, 50, 154
Derman, C., 32, 33, 62, 83, 102, 116, 117, 137, 138, 154
Deterministic (time invariant) policies, 7
Discounted cost
   criterion, 6
   problem, 6
Dodge, H. F., 138, 155
Doob, J. L., 116, 155
Dynamic programming, 11
   functional equations of, 14
Dynkin, E. B., 116, 118, 155

E

Eaton, J. H., 62, 155
Entrance fees, 117
Ergodic class, 140

F

Feller, W., 142, 147, 155
Fisher, L., 33, 155
Fox, B. L., 84, 154

G

Gillette, D., 33, 155

H

History, 3

Horizon, 5
   finite horizon problem, 11
Howard, R. A., 8, 50, 83, 155

I

Inventory system under periodic review, 4
Irreducible state space, 78, 140

K

Karlin, S., 32, 152, 155
Kemeny, J. G., 142, 155
Klein, M., 62, 138, 154, 155
Kolesar, P., 137, 155

L

LaSalle, J. P., 8
Laws of motion, 1, 3
Lieberman, C. J., 138, 155
Liggett, T. M., 33, 155
Linear programming formulation of
   average cost problem, 73
   discounted cost problem, 41
   first-passage problem, 57
   stopping problem, 109, 113
Linear programming problem, 150
   dual, 151
   feasible solution of, 150
   objective function of, 150
   optimal solution of, 150
   primal, 151
   simplex method solution, of, 150
Lippman, S., 33, 155
Loeve, M., 147, 155

M

Maitra, A., 49, 155
Manne, A. S., 8, 84, 155
Markov chains with stationary transition probabilities, 139

Index

Markovian decision model, 2
Markovian decision process, 1, 4
Mean first-passage time, 142
Metric, 144
Metric space, 144
Miller, B. L., 84, 155

O

Optimal first-passage problem, 5
Optimality principle, 15

P

Policy, 3
  convergence of policies, 20
  deterministic, 7
  memoryless (Markovian), 6
  renewal, 90
  time invariant (Markovian), 7
Policy improvement interation for
  average cost problem, 71
  discounted cost problem, 40
  first-passage problem, 56
Policy improvement procedure for
  average cost problem, 71
  discounted cost problem, 41
  first-passage problem, 56
Pursuit problem, 62

R

Randomization, 3
Recurrent event process, 90, 146
Recurrent states, 141
Replacement model, 121
Robbins, H., 116, 117, 154
Ross, S., 33, 155
Rule, *see* Policy

S

Sacks, J. S., 116, 117, 154
Sequential search problem, 132

Shapley, L. S., 8, 49, 155
Snell, J. L., 116, 142, 155, 156
State-action frequencies, 98
  expected, 89
States, 2
  sequence of 2
Stochastic games, 8
Stochastic traveling salesman problem, 135
Stopped process, 104
Stopping time, 104
Strauch, R., 102, 154, 156
Successive approximations
  for discounted cost problem, 36
  for first-passage problem, 54
  for stopping problem, 109
Super-regular function, 106
  smallest, dominating a function, 108
Surveillance-maintenance-replacement model, 127

T

Target state, 5
Taylor, H., 116, 156
Transient states, 141
Transition probabilities, 139
  t-step, 140
Tychonov's theorem, 20, 32, 147

V

Veinott, A. F. Jr., 33, 62, 83, 84, 102, 154, 156

W

White, L. S., 138, 156
Widder, D., 147, 156

Z

Zadeh, L. A., 62, 155

Mathematics in Science and Engineering

A Series of Monographs and Textbooks
Edited by RICHARD BELLMAN, *University of Southern California*

1. T. Y. Thomas. Concepts from Tensor Analysis and Differential Geometry. Second Edition. 1965
2. T. Y. Thomas. Plastic Flow and Fracture in Solids. 1961
3. R. Aris. The Optimal Design of Chemical Reactors: A Study in Dynamic Programming. 1961
4. J. LaSalle and S. Lefschetz. Stability by by Liapunov's Direct Method with Applications. 1961
5. G. Leitmann (ed.). Optimization Techniques: With Applications to Aerospace Systems. 1962
6. R. Bellman and K. L. Cooke. Differential-Difference Equations. 1963
7. F. A. Haight. Mathematical Theories of Traffic Flow. 1963
8. F. V. Atkinson. Discrete and Continuous Boundary Problems. 1964
9. A. Jeffrey and T. Taniuti. Non-Linear Wave Propagation: With Applications to Physics and Magnetohydrodynamics. 1964
10. J. T. Tou. Optimum Design of Digital Control Systems. 1963.
11. H. Flanders. Differential Forms: With Applications to the Physical Sciences. 1963
12. S. M. Roberts. Dynamic Programming in Chemical Engineering and Process Control. 1964
13. S. Lefschetz. Stability of Nonlinear Control Systems. 1965
14. D. N. Chorafas. Systems and Simulation. 1965
15. A. A. Pervozvanskii. Random Processes in Nonlinear Control Systems. 1965
16. M. C. Pease, III. Methods of Matrix Algebra. 1965
17. V. E. Benes. Mathematical Theory of Connecting Networks and Telephone Traffic. 1965
18. W. F. Ames. Nonlinear Partial Differential Equations in Engineering. 1965
19. J. Aczel. Lectures on Functional Equations and Their Applications. 1966
20. R. E. Murphy. Adaptive Processes in Economic Systems. 1965
21. S. E. Dreyfus. Dynamic Programming and the Calculus of Variations. 1965
22. A. A. Fel'dbaum. Optimal Control Systems. 1965
23. A. Halanay. Differential Equations: Stability, Oscillations, Time Lags. 1966
24. M. N. Oguztoreli. Time-Lag Control Systems. 1966
25. D. Sworder. Optimal Adaptive Control Systems. 1966
26. M. Ash. Optimal Shutdown Control of Nuclear Reactors. 1966
27. D. N. Chorafas. Control System Functions and Programming Approaches (In Two Volumes). 1966
28. N. P. Erugin. Linear Systems of Ordinary Differential Equations. 1966
29. S. Marcus. Algebraic Linguistics; Analytical Models. 1967
30. A. M. Liapunov. Stability of Motion. 1966
31. G. Leitmann (ed.). Topics in Optimization. 1967
32. M. Aoki. Optimization of Stochastic Systems. 1967
33. H. J. Kushner. Stochastic Stability and control. 1967
34. M. Urabe. Nonlinear Autonomous Oscillations. 1967
35. F. Calogero. Variable Phase Approach to Potential Scattering. 1967
36. A. Kaufmann. Graphs, Dynamic Programming, and Finite Games. 1967
37. A. Kaufmann and R. Cruon. Dynamic Programming: Sequential Scientific Management. 1967
38. J. H. Ahlberg, E. N. Nilson, and J. L. Walsh. The Theory of Splines and Their Applications. 1967

39. Y. Sawaragi, Y. Sunahara, and T. Nakamizo. Statistical Decision Theory in Adaptive Control Systems. 1967

40. R. Bellman. Introduction to the Mathematical Theory of Control Processes Volume I. 1967 (Volumes II and III in preparation)

41. E. S. Lee. Quasilinearization and Invariant Imbedding. 1968

42. W. Ames. Nonlinear Ordinary Differential Equations in Transport Processes. 1968

43. W. Miller, Jr. Lie Theory and Special Functions. 1968

44. P. B. Bailey, L. F. Shampine, and P. E. Waltman. Nonlinear Two Point Boundary Value Problems. 1968.

45. Iu. P. Petrov. Variational Methods in Optimum Control Theory. 1968

46. O. A. Ladyzhenskaya and N. N. Ural'tseva. Linear and Quasilinear Elliptic Equations. 1968

47. A. Kaufmann and R. Faure. Introduction to Operations Research. 1968

48. C. A. Swanson. Comparison and Oscillation Theory of Linear Differential Equations. 1968

49. R. Hermann. Differential Geometry and the Calculus of Variations. 1968

50. N. K. Jaiswal. Priority Queues. 1968

51. H. Nikaido. Convex Structures and Economic Theory. 1968

52. K. S. Fu. Sequential Methods in Pattern Recognition and Machine Learning. 1968

53. Y. L. Luke. The Special Functions and Their Approximations (In Two Volumes). 1969

54. R. P. Gilbert. Function Theoretic Methods in Partial Differential Equations. 1969

55. V. Lakshmikantham and S. Leela. Differential and Integral Inequalities (In Two Volumes). 1969

56. S. H. Hermes and J. P. LaSalle. Functional Analysis and Time Optimal Control. 1969.

57. M. Iri. Network Flow, Transportation, and Scheduling: Theory and Algorithms. 1969

58. A. Blaquiere, F. Gerard, and G. Leitmann. Quantitative and Qualitative Games. 1969

59. P. L. Falb and J. L. de Jong. Successive Approximation Methods in Control and Oscillation Theory. 1969

60. G. Rosen. Formulations of Classical and Quantum Dynamical Theory. 1969

61. R. Bellman. Methods of Nonlinear Analysis, Volume I. 1970

62. R. Bellman, K. L. Cooke, and J. A. Lockett. Algorithms, Graphs, and Computers. 1970

63. E. J. Beltrami. An Algorithmic Approach to Nonlinear Analysis and Optimization. 1970

64. A. H. Jazwinski. Stochastic Processes and Filtering Theory. 1970

65. P. Dyer and S. R. McReynolds. The Computation and Theory of Optimal Control, 1970

66. J. M. Mendel and K. S. Fu (eds.). Adaptive, Learning, and Pattern Recognition Systems: Theory and Applications, 1970

67. C. Derman. Finite State Markovian Decision Processes, 1970

68. M. Mesarovic, D. Macko, and Y. Takahara. Theory of Hierarchical Multilevel Systems, 1970

69. H. H. Happ. Diakoptics and Networks, 1970

In preparation

G. A. Baker, Jr. and J. L. Gammel, eds. The Padé Approximant in Theoretical Physics

Karl Astrom. Introduction to Stochastic Control Theory

C. Berge. Principles of Combinatorics

Ya. Z. Tsypkin. Adaptation and Learning in Automatic Systems

T
57.83
D46

DEC 15 1971